Editors:

G.E. Blight
Department of Civil Enginee
Johannesburg, South Africa

E.C. Leong
Nanyang Technological Univ

 **CRC Press**
Taylor & Francis Group
Boca Raton London New York Lei

CRC Press is an imprint of the
Taylor & Francis Group, an **informa** business

A BALKEMA BOOK

First issued in paperback 2017

CRC Press/Balkema is an imprint of an informa business

© 2012 Taylor & Francis Group, L

Typeset by MPS Limited, Chennai,

Library of Congress Cataloging-in-Pul

Mechanics of residual soils / editor
      p. cm.
   "A Balkema book."
   Includes bibliographical referenc(
   ISBN 978-0-415-62120-5 (hardba
   1. Soil mechanics.   2. Residual ι
4. Soils—Tropics.   I. Blight, G. E.
   TA709.5. M4183   2012
   624.1'5136—dc23

Published by: CRC Press/Balkema
                P.O. Box 447, 2300 ,
                e-mail: Pub.NL@tayl
                www.crcpress.com –

ISBN 13: 978-1-138-07224-4 (pbk)
ISBN 13: 978-0-415-62120-5 (hbk)

Preface to second edition
Acknowledgements
Author biographies
List of abbreviations and m

1 Origin and formation of re
G.E. Blight

    1.1   Definitions relating to
    1.2   Rock weathering proc
    1.3   The effects of climate
    1.4   The effects of topogra
    1.5   General characteristic
    1.6   Depth and intensity o
    1.7   More about pedocret
    1.8   Transported soils that
    1.9   The weathering of sol
    1.10 Dispersive soils
    1.11 Relict structures in re
           performance
    1.12 Rapidity of weatherir
    1.13 Detailed examination
           igneous rock
    1.14 An introduction to th
   References

2 Microstructure, mineralog
A.B. Fourie, T.Y. Irfan, J.B. Queiroz

    2.1   Microstructure and m
    2.2   Mineralogy and occu
    2.3   Determination of min
    2.4   Microstructure of res
    2.5   Mineralogy and micrc
    2.6   Examples of the mine

3.2   Soil engineering dat;
3.3   Detailed informatior
3.4   Profile description
3.5   Simple *in situ* tests a
3.6   Taking undisturbed
References

4   The mechanics of compa
G.E. Blight & J.V. Simmons

4.1   The compaction pro
4.2   Consequences of un
4.3   The mechanisms of
4.4   Laboratory compact
4.5   Precautions to be ta
4.6   Roller compaction i
4.7   Relationships betwe
       and optimum water
4.8   Designing a compac
4.9   Seepage through fiel
4.10   Control of compacti
4.11   Special consideratio
       of evaporation
4.12   Additional points fo
4.13   Compaction of resic
4.14   The mechanics of ur
       after construction
4.15   Pore air pressures ca
4.16   Summary
References

5   Steady and unsteady flov
     soils – permeability of sa
V.K. Garga & G.E. Blight

5.1   Darcy's and Fick's la
5.2   Displacement of wai

6.12   Permeability characte...

References

6   Compressibility, settleme...
R.D. Barksdale & G.E. Blight

6.1   Compressibility of res...
6.2   The process of compr...
6.3   Biotic activity
6.4   Measuring the compr...
6.5   Settlement prediction
      foundations
6.6   Settlement prediction...
6.7   Movement of shallow...
6.8   Collapse of residual s...
References

7   Shear strength behaviour
strength in residual soils
R.P. Brenner, V.K. Garga & G.E. ...

7.1   Behaviour and differe...
      transported soils
7.2   Laboratory strength t...
7.3   Field strength testing
References

8   Case histories involving v...
of residual soils
G.E. Blight

8.1   Settlement of two tow...
8.2   Settlement of an earth...
      residual soil
8.3   Settlement of an apar...
8.4   Preheaving of expans...
8.5   Heave analysis for a ...
      at an experimental sit...

The first edition of "Mechan
A.A. Balkema, whose successor;
bers of Technical Committee 2.
Foundation Engineering and edi
Committee 25.

When, 15 years later, the nee(
was not considered feasible to a;
moved on to other interests, o
that subsequent editions of a l
decided that the chapters woul
original authors, but the conter
approach to the content of the b
E.C. Leong to co-author/edit th

The main changes and exten;

All of the diagrams have beei
being discovered and corrected
of soils, is now not only illustrate
that are grouped with a further §
new topics have been introduce
passing, have been amplified. A

- a substantial section on ped
- consideration of the weathe
- dispersive soils and their ad
- an introduction to the mecl

To eliminate repetition that
have been consolidated into a
classification of residual soils. T
the original five authors.

Chapter 3, on soil profile
illustrated by a colour reproduc

Chapter 4, on compaction,
permeability of compacted so
required maximum water-perm

appendix is now integrated in...

Chapter 7, on shear stren
Chapter 9, with the addition
of the mechanics of shearing ii

The final Chapter 8 (forme
tories of settlement and heave
of an earth dam embankment
the effects of heave by in situ
tures incorporating compacted
residual soils have also been ir

Overall, the content of Edit
Edition 1.

Geoff Blight,
Johannesburg,
2012

We, the editors/authors of the
original "Mechanics of Residu;
V.K. Garga, T.Y. Irfan, J.B. Q
Without their pioneering effort
edition of this book.

We also thank all the geotec
continued, in the 15 years since t
soils and to pool their knowled;

Geoff Blight has had the goo
in the field of residual soils who
Africa. He wishes to acknowled
soils and his growth of knowlec

Gordon Aitchison, George A
Brand, Peter Brenner, Tony Bri
Lou Collins, George Dehlen, (
Fourie, Vinod Garga, Malcolm
Lumb, Ken Lyell, Dirk van der .
Philip Paige-Green, Tim Partri
Harianto Rahardjo, Brian Richa
Ken Schwartz, George Sowers, 1
Harold Weber, Laurie Wesley, A

Geoff Blight thanks them, or
residual soils.

E.C. Leong thanks Geoff Blig
It was a privilege and joy to wo

Cathy Snow prepared all of tl
additions and deletions that we.

We both acknowledge and
support.

Geoff particularly thanks hi
another major manuscript.

Unless otherwise acknowled;

Geoffrey
at the U
his PhD
some of t
soils, und
early wor
1963, pr
tions of
became i
on unsatu
the Interr
Engineering's Technical Commi
from 1982 to 1997 and serve
co-authored the first edition of
during his Chairmanship.

He has also authored or co-au
bulk storage structures" (2006
facilities" (2010) and "Alkali-a
(2011), all published by CRC P

Eng-Cho
of Civil
Singapore
National
Western /
Nanyang
ests in un
Harianto
dynamics
has publis
He has ex
urated and unsaturated soils an
developed a number of specializ
systems for field applications.
cal committees on standards. H
accreditation and SPRING, the

(In many cases, the same symbc
to tell from the context which
used.)

ROMAN LETTERS

A	Skempton-Bishop
A	cross-sectional ar
A	$(\tan \theta - \sec \theta + 1)$
AASHTO	American Associa
	Transportation O
Al_2O_3	alumina
AMSL	Above Mean Sea
B	Skempton-Bishop
c	cohesion in total
c'	cohesion in effect
c_v	coefficient of con
C	compressibility
C_c, C_r	compression and
$CaCO_3$	calcium carbonat
$Ca(HCO_3)_2$	calcium bicarbon
CD	consolidated, dra
CEC	Cation Exchange
CPT	cone penetration
CU	consolidated, unc
d	depth
D	depth
D_c	diffusion coefficie
D_e	Entrance diamete
D_i	Internal diameter
D_w	Diameter of wall
e	void ratio
E_I	elastic modulus
E_h	elastic modulus n
Ej	pan evaporation

I_f	influence fac...
k	coefficient of
k_h, k_v	coefficients o... or vertical fl...
K	stress ratio
K_A, K_o, K_p	active, at rest
LL	liquid limit
LVDT	Linear Volta...
m	mass of air s...
m_a	molecular m...
m_s, m_w	number of m... (eqn 1.10)
M	mass of air e...
$MgCO_3$	magnesium c...
$Mg(HCO_3)_2$	magnesium b...
n	porosity
n_a	air porosity
n_w	water porosi...
N	Constant use
N	Number of b... Test (SPT)
N_c	bearing capa...
$N = 12Ej/Pa$	Weinert's N
p, P	pore water s...
$p' = \frac{1}{2}(\sigma_1' + \sigma_3')$	mean effectiv...
p''	pore water s...
psd	particle size ...
Pa	annual rain ...
P, Pav	Pressure, ave...
PI	plasticity ind...
PL	plastic limit
P_L	limit pressur...
q	applied stres...
q	quantity of fl...
$q' = \frac{1}{2}(\sigma_1 - \sigma_3)$	maximum sh...
r	radius of liqu...

S.O₂	silica
SWCC	Suction Water Content Characteristic Curve (s
t, t$_f$	time, time to failure
T	basic time lag
T	surface tension of a liq
T	time factor in consolid
T	Ton
T	Torque
TDS	Total Dissolved Solids
TEM	Transmission Electron
u	pore pressure
u$_a$	pore air pressure
u$_w$	pore water pressure
U	degree of consolidatio
UU	unconsolidated, undra
USCS	Unified Soil Classificat Casagrande classificati
v	flow velocity
V	volume
VCL	Virgin Compression L
w	gravimetric water cont
XRD	X-Ray Diffraction

GREEK LETTERS

α	alpha	rheological factor page 185 or angle
γ	gamma	bulk or total unit
γ$_d$		dry unit weight
γ$_{sat}$		saturated unit wei
δ	delta	displacement
ε	epsilon	linear strain
ε$_v$		volumetric strain (eqn. 6.3 (page 156
ε$_1$, ε$_2$, ε$_3$		principal strains

σ'		effective di
$\sigma_1, \sigma_2, \sigma_3$		principal st
σ_g		intergranul
τ, τ_f	tau	shear stress
φ	phi	angle of sh
		[°deg. of ar
φ'		angle of sh
		stresses
φ^b		Fredlund's
		increases w
χ	chi	Bishop's eff
χ_s, χ_m		Bishop's pa
		matrix suct

OTHER SYMBOLS

∂ partial differential operator

SI AND RELATED UNITS

kg	kilogram	
N	Newton	[1 N = 1 kg
J	Joule	[1 J = 1 Nm
Pa	Pascal	[1 Pa = 1 N
bar		[1 bar or a
s	second	
h	hour	
d	day	
y	year	
Note:	the unit for electrical c	

1.1 DEFINITIONS RELAT

The definition of a residual soil v
definition would be:

> "A residual soil is a soil-li
> decomposition of rock or rc
> original location".

In this definition, "rock" re
to materials such as grains of a
be a continuous gradation from
through weathered soft rock ar
product of decomposition of the
secondary deposits of alumina
resemblance to the parent mate:

The differences between tr
deposits are illustrated by Figi
Figure 1.1a represents the void
tion. V_v and V_s are the void anc
is high and the effective overbi
increases with time (Figure 1.1
burden is higher, the void ratio
the parent rock of the residual
sist of small occluded spaces. /
progress, the solids volume decr
of small detached weathered so
on a micro-scale). Simultaneou:
overburden stress reduces. This
and accelerates the weathering [
has developed a system of inter-
meable to penetration by air a
(evaporation at the surface exc
be drawn to the surface and wi

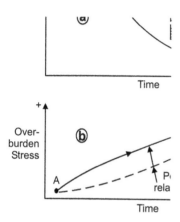

Over-
burden
Stress

A

P
rela

Time

Figure 1.1 Basic differences in the fo
2010).

carbonate, calcium sulphate o
decrease along E to D (in Figt
cases causing a cementing of tl
  Residual soils can have ch
transported soils. For exampl
size is inapplicable to many r
aggregates of particles or wea
become progressively finer if 1
What appears *in situ* to be a c
during excavation, mixing and
usually be related to its granu
the reason just explained, this
of residual soils is usually gov
by relict or superimposed featu
Similar remarks apply to prop
The value of G_s, for example
measured has been comminut
higher values of G_s.

....

Biological weathering includes
and chemical action (e.g., bacte
and sulphur compounds (e.g., P

Most commonly, residual s
but residual soils formed from
et al., 2002) Chemical process
rocks and metamorphic rocks
processes dominate the weathe
nating from sedimentary rocks
closely interrelated that one pro
other.

Occasionally, residual soils
sediments. The commonest exam
weathering of feldspars in depo
Schwartz & Yates, 1980). Shira
is an unconsolidated volcanic s
properties have much in commo
humid tropical climates, fluvial
flood plains and deltas are ofte
2001; Zhang et al., 2007), trans
canic ash, deposited aerially or
and other clastic material that h
(e.g., Wesley, 2010).

Hydrolysis is considered to
processes (Zaruba & Mencl, 19
an acid and a base. In rock wea
the reaction is a clay mineral. C
rocks containing iron sulphates
usually have a larger specific vo
to oxidation increases the void
rock (e.g., Mason 1949).

The breakdown of one clay
of ions between percolating solu
and calcium are the most readi
basic structure of the clay miner
converting, for example, an illi

Trop

Figure 1.2 Suggested influence c

calcium bentonite. Although t
physical properties may suffer
  Bacteria may play the role
the oxidation of sulphide mine
activity of the bacteria thio-ba
  Chelation is a process whe
of hydrolysis. Jackson & Kell
basalt surfaces is greater and
cover than if lichen is not pres

1.3 THE EFFECTS OF CI

Climate exerts a considerable i
1964 and 1974; Morin & Aye
dry climates while the extent a
the availability of moisture an
reaction rates approximately c
  According to Uehara (198
in a predictable way with dista
a gross over-simplification bec
formly with distance from the
upper air streams and ocean ci
cept of the influence of climat
high temperatures and year-rc
low activity kaolin and oxid

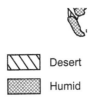

| ▨ | Desert |
| ▩ | Humid |

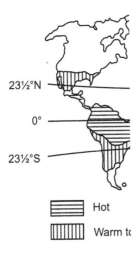

23½°N

0°

23½°S

| ▤ | Hot |
| ▥ | Warm to |

Figure 1.3 Climatic zones of the worl
climates.

Activity = Plasticity Index/% pa
limits of the tropics and high act
1.5 to 7). The above concept h
(1967), whose diagram (Figure
weathering and the formation of
etc. The influence of temperatu
has been correlated with Weine

$$N = \frac{12Ej}{Pa}$$

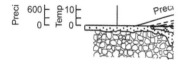

Moisture and temperature
both diminish towards the
poles; weathering and
organic matter decomposition
are slow, and low precipitation
(polar desert) or permafrost
(tundra) inhibit leaching of
mobile constituents.

Hig
higl
ten
slov
dec
wea

Figure 1.4 Influence of global clima
1967).

Where Ej = Potential evaporati
southern hemisphere month, a

A value of N = 5 marks th
chemical weathering predomi
physical weathering is the mor
thicknesses of residual soil m
greater. Where N exceeds 5, t
the degree of weathering is les
by Weinert's N = 5 contour. I
predominates, while to the we
map also indicates how clima
Mozambique current flows sc
warm humid sub-tropical clim
west coast, resulting in a cool

Climate has a further pos;
that of unsaturation. Even in st
often deeper than 5 to 10 m an

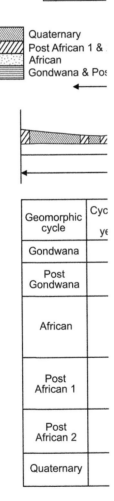

B
C

Quaternary
Post African 1 &
African
Gondwana & Pos

| Geomorphic cycle | Cyc
ye |
|---|---|
| Gondwana | |
| Post Gondwana | |
| African | |
| Post African 1 | |
| Post African 2 | |
| Quaternary | |

Figure 1.5 (Above): Subdivision of So
section AA showing sub-c
surfaces, as well as ages of

The effective stress σ' is gove
$(u_a - u_w)$ where σ is the appli
air and pore water pressures
the effectiveness of $(u_a - u_w)$
situations, u_a equals the atm
conventional form of the effec

$$\sigma' = \sigma - u_w$$

can be used with little error f
saturation. In an unsaturated s
water stress is added to the to
is at a depth of 10 m, this ad
soil profile and about 50 kPa t
be lost during periods of prol
conditions of unsaturation are
unsaturated soils are discussed

1.4 THE EFFECTS OF T(

For a deep residual soil profile
the earth's crust must exceed
erosion.

Topography controls the
of available water and the ra
Precipitation will run off hills
controls the effective age of the
material from the surface. Th
valleys and on gentle slopes ra
Ayetey, 1971). At least part of
consist of colluvium eroded fi
engineer must distinguish betw
behaviour may be quite differe

It thus follows that ancien
mation than does more recen

of South Africa. The residual so
characteristics owing to the grea
Because the surfaces are so old,
deep (e.g., Figure 1.13).

In the Johannesburg area t
erosion has formed several Post
that has suffered widespread lo
the 'Post African 1 and 2 surface
of the African surface also occu
since 'Africa' times. This zone
undulating topography, and is
distinct surfaces or zones are c
Above African to the African to
1.6 later, there are distinct diff
underlies each surface. The topc
locations of the ancient erosion
together with a tabulation of th
problems associated with soils l

Fitzpatrick & Le Roux (19
rock on hillsides beneath the A
weathering increased down the
dominant clay minerals at the t
the slope was smectite. This is a
soils probably include a certain
deposited lower down the slop
between transported and residu

van der Merwe (1965) mac
basic igneous rock types. In a
ferent sites he found that the p
affecting development of clay n
Samples taken from a site hig
and vermiculite to be the domi
vermiculite, montmorillonite ai
flat, with impeded drainage. Be
presence of predominantly redu
kaolinite stage and hence mont
study indicated that good inter

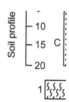

Figure 1.6 Section of black clay top
consists of grassy meado
ture (breaking down to cr
structure (breaking down
enslides); 3. As for 2, but
(approximately 60 mm) o
5. Pleyed clay with slicker

development of kaolinite whils
of montmorillonite.

Topography also controls
influence the amount of moistu
table, and thereby the depth (
The relief of the area thus con
or good drainage.

Under conditions of good
an essential mechanism of roc
instance are found to be more
the annual precipitation and
Figure 1.7 which shows some
annual rainfall for weathered ¿
on current rainfall patterns an(
times. Obviously, the higher t
leaching will occur, and the hi¿

1.5 GENERAL CHARAC

(Illustrated by colour plates (C
The process of formation (
difficult to understand and dif

0.8

0.6
400   600

Figure 1.7 Relationship between void
in South Africa, under the

It is evident that apart fro
the properties of a residual soi
individual consideration and it i
area to predict conditions in an
two areas is similar. For instance
Malaysian peninsular may have
semi-arid Highveld plateau in S
south western coast of South A
decomposition results from the
recent past and present.

The chemical changes and
extremely complex. For example
mation of clay minerals is show
may be arrested at any stage an
result of changes in climate or c

Figure 1.8b (Gonzalez de
volcanic rock observed in Camei
unknown differences in mineral
a lower altitude and hotter, mois
the difference in weathering seq

As mentioned previously, a
has shown that on the South
well drained situations over no
develop from identical parent ro
when flying over or even drivin
been stripped of their crops ar

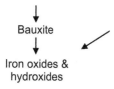

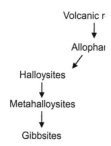

Figure 1.8 (a) Suggested sequence o
1965). (b) Sequences in tl
of Africa. (Gonzalez de V

surfaces of rises and hillsides ar
black.

   Wesley (1973) concluded
in Java originate from much th
different ages.

   Because weathering proce
faces and other percolation pa
increasing depth and reducing
faces. In profiles residual from
rock are very often found enclo
This is the typical "onion skin"
such as basalts and dolerites. '
indistinctly divided zones (Va
1969) as illustrated in Figure

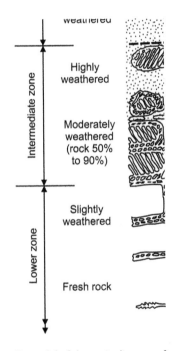

weathered

Highly
weathered

Intermediate zone

Moderately
weathered
(rock 50%
to 90%)

Slightly
weathered

Lower zone

Fresh rock

Figure 1.9 Schematic diagram of

and leached soil often reworkec
and intersected by root channels
transport processes. The interm
but exhibits some features of tl
stones. This zone often contains
salts which may give it a mottle

Colour Plates C1 and C2 (
shown in Figure 1.9. The profi
shows the maximum degree of v
transported silty, sandy soil, in v
rock, overlying *in situ* weather
field of view shows what was

.... amygdales.

Plate C3 shows a residual
to a reddish-brown sandy silt.
The boulders in the centre-rig
dyke, which dips vertically.

Plate C4 shows a weathere
and containing core boulders
boulders are particularly inter
entrained by the flowing lava
from the surface over which it

Plate C5 shows an andesit
pletely weathered to a great dej
horizontally are traction wires
reddish-coloured residual ande
marked by a line of light-colou
often occurs near the top surfa
of an ancient erosion surface.

Saprolites are materials tha
ified but recognizable relics of
example a saprolite derived fro
and amygdales of the parent ro
and jointing pattern of its pare
depth.

Plate C6 shows a profile of
The inclined lines of coarse pa
the ash layer was built up by sud
left of the photo, a displaceme
been faulted, with the fault rui

Laterites are usually high
that contain a sufficient conce
have been cemented to some de
from the evaporation of iron-
in Figure 1.1c). Depending on
described as lateritic or as late
ancient transported soils may
nance of either iron or alumin
as ferricretes or alucretes. Desa

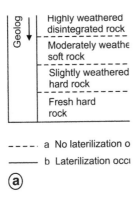

Figure 1.10 (a) Changes occurring in Sueoka, 1988).

in terms of the silica/alumina rat
For lateritic soils SiO_2/Al_2O_3 lie
1.3. Madu (1980) shows that ch
be related to the strength of the

The above remarks apply e
of calcium carbonates, in whic
other cases silica can be precipi
term "pedocrete" is used to des
been cemented by secondary de

Figure 1.10a (adapted from
schematic form the progression
the progression from residual so
not inevitable, but depends on
for precipitation of the salts to
sion (Futai *et al.*, 2004) in a pro
saprolitic soil between 2 and 7 r
any cementing, but Figure 2.4b
amorphous iron salts as the so
saturated which is probably wh

Most of the depth of a resid
tion. Because of their mode of fo

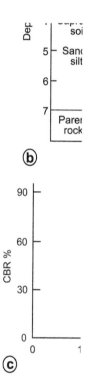

Figure 1.10 (b) Weathering profile (
content and CBR (Madu

strata of limited depth. Calcre
rials for the building of roads
1988; Sweere *et al.*, 1988; Gid
for this purpose. They may al
strength and low compressibili
properties indicated in Figure 1
(Madu, 1980).

Plate C7 shows a layer of
The small, spherical pebbles co
core stones from the residual s

As stated earlier, the formation o
on the rate of erosion of the surf
the surface equals or exceeds th
rock, no residual soil will form
in a relatively small area over
weathered profiles will occur be
indeed so for andesite profiles. I
can be summarized as follows:

- The total thickness of residu
  more than 100 million year
  to that beneath the less than
  averages only 16 m).
- The predominant soil colo
  profiles it is yellow.

Plate C4 illustrates a lighter
red, older profile.

As mentioned in section 1.
in the weathering of sedimenta
weathering of igneous rocks. H
a profile through a diabase sill
small angular boulders with onl
by Plate C3. The present rainfa
the site of Plate C3 and the ten
can be that the sill of Plate C9 v
However, igneous rocks can an
Plate C10, for example is an e
granite has deeply weathered to

Sedimentary rocks can also
shows shallow weathering of a
rock at the base of the exposure
shows an exposure of weathere
the overlying transported layer
stiff fissured clay.

As discussed earlier in conj
at a particular depth. Plate C1

(....). ....
he has collected over the past ·

Occurrence

Pedocretes are believed to forn
mates such as those defined in .
in areas where 12 times the ma
by the annual precipitation is
Local drainage conditions mus
charged with potentially pedo·
for the salts to precipitate in th
water-courses, pans or depress
out in the dry season.

Obviously, the type of sal
be available in the underlying
occurrences of a particular ty
contour in Figure 1.5, along th
mesa cappings, associated witl
Inland, silcretes are associated
Gypcretes occur mostly in the

Calcretes occur in somewl
supplying calcium- and magn
calcrete zone. For example, in .
the excavation and occurs abou
commonly associated with sea
water rising in the rainy seasc
dry season.

Formation and form

The rate of formation of pedoc
20 and 200 mm thickness per :
1898, during the South Africa
of shell cases and lead slugs :
in ferricrete with layer thickn
postulated rate of deposition of
of over 30 m, but are seldom

NP = non plastic
Casagrande classification – see Figure 2.11
10% FACT = 10% Fines Aggregate Crushin
Moh hardness = hardness of calcrete parti

similar thicknesses occurs bene
its equivalent in Australia. Cal
thicknesses of 10 m in places.

Pedocretes occur in a numb
as construction materials are lis

- Calcareous, ferruginized or
  particle sizes that are rich
  cementation or nodule deve
- Powder calcrete, ferricrete (
  or aggregates of particles of
- Nodular pedocretes consist
  ferruginous or siliceous par
- Honeycomb pedocretes in ·
  other to form a coherent sk
- Hardpan pedocretes are ind
  from soft to very hard roc
  usually overlie less cemente

Engineering properties

Some engineering properties rel
rized in Table 1.1.

Particle specific gravity G

Because pedocrete grains or noc
difficult to define and measure. I
G for calcretes falls within the ra
on the content of iron salts and

Foundations on pedocretes

As mentioned above, the quali
deposit and usually lessens wi
and foundation exploration sl
well as the usual test-pitting ar
have collapsible grain structur
in situ and lose strength when
of as much as 90% have been

Small scale karst features a
have also been found in lateri
weathered calcretes.

Pedocretes in road constru

The widest recorded engineerir
whole of southern Africa, silcro
unpaved roads and for all laye
surfaced roads. Similar uses ar
and Brazil. An interesting var
water as the wearing course o
roads are periodically watered
existent, sea fogs ensure that
water, absorbed by the high sa
road surface. The surface beco
roads look superficially like co
Plate C23).

In general, pedocretes po
expected from their gradings
if they develop, it is usually ir
stands.

Pedocretes as building bloc

Laterite has been used for mi
blocks. Both laterite and calic
deposits and allowed to dry i

dissolve and have also been k
structures or structures to reta
proceeded with after careful ma

1.8 TRANSPORTED SOIL

It was mentioned in section 1.1
and aeolian sands may weathe
already been described by Plate
Pinatuba on Luzon island in the
quantities of ash that blanketed
most productive rice fields in the
recorded since 1941, with the h
the years following the eruptio
flows that travelled vast distanc
or changing the course of rivers
a newly cut meander in a river
be discerned on the far horizor
through grey ash from the mos
buried village that had been bu
lower stratum of ash can be seer
Hence deposits of this sort, inclu
of alternating deposition and w
defined separate and successive
is highly fertile, the new ash th
However, the tall growth of gras
probably be short-lived, and it
years.

1.9 THE WEATHERING O
AND DOLOMITES

The main agent of weathering o
rain water infiltrating the surfac

dolomite can remove as much
million L or kg) of percolating
can be formed. This means th
rate of 100 ML per day could
dolomitic "eyes" with flows as
or limestone terrain.

Figure 1.11 illustrates the
of near-surface rock pinnacle
dolomite, with the residuum
residuum is often very rich in
ganese by excavating and proc
pillars (e.g., Viljoen & Reimol

Two types of geotechnical
by Figure 1.11 or Plate C15, (
into sink-holes.

- The residuum between so
  texture shown in Plate C1
  are not designed to span b
  building they support. Pla
  cause.
- The second problem is the
  lapse of the residuum into
  terrain can vary from 5 m

Plate 1.1 is a photograph
more than 30 m deep that eng
mine in 1962. The accident ha
when the plant was fully opera
the plant components or bodi
sink-hole, also measuring mor
and a family of 5, also withou
4 houses on the site shown i
houses were awoken by the no
the second house fell into the
the remaining houses also disa

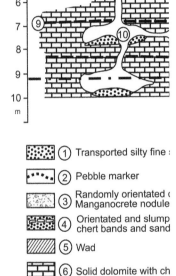

1. Transported silty fine :
2. Pebble marker
3. Randomly orientated (
   Manganocrete nodule
4. Orientated and slump
   chert bands and sand
5. Wad
6. Solid dolomite with ch

Figure 1.11 **The process of weatheri
solid rock with soft resid**

bodies of victims were recovere
now-defunct newspapers of the

Sink-holes are an ever-pres(
naturally, but often their occurre
to pumping water from the aqu
By removing buoyancy from th
the self-weight of the drained str
and rock rubble arches and the s
both as shown in Figure 1.11. I
scale by pumping water from a d
natural condition of soil strata
reduces the strength of the soil b
term desiccation, and simultan(
unit weight, both contributing t

Plate 1.1   Sink hole that er

Plate 1.2

more readily than others, and ai
in certain soils, both transporte
disperse when exposed to wate
dissolved solid content. This is
repulsive electrical surface char
than the attractive van der Waa

When this happens, the exti
the soil matrix, as if in solution
result is that internal erosion is
This can cause the formation of
collapse of earth structures, es|
Elges, 1985; Paige-Green, 2009

Since problems with disper
geotechnical literature (e.g., Ait
been devised. These include the:

- Pinhole test, in which erodil
  pierced by means of a hyp
  et al., 1976).
- Cation exchange capacity
  plot, in which these two m
  1987).
- Crumb test, in which a crui
  to stand and observed to se
  still water around the crum
- Sodium adsorption ratio (S.
- Total dissolved solids (TDS

Bell & Walker (2000) carrie
came to the conclusions summa
identifying dispersive soils avail

As is the case with other pi
ter 4) both of the physical tests i
performed on soil as close to its
drying, let alone oven dryness v

Also concerning residual so
are low in calcium and magnes

sodium chemistry in Table 1..
dispersive, as have soils residu
been found to be dispersive. In
or even dispersive is the preser
bad-lands of South Dakota in
in Turkish Cappadocia (see Pl
the rapid rate of surface eros
occasionally interrupted by ve:
rock.

Physical problems with di
persive tendencies, seepage wa
that of the *in situ* pore water, a
of the soil in contact with a r
the side or base of a backfillec
or along the side of an outlet
piping between the concrete o1
dam constructed of compactec
tendencies, but the main reaso
probably arching of the fill ov
against the sides of the condu
against the tower and outlet cc
possibly, the low dissolved sol:
list, poor compaction and red
the most important. (See also :

1.11 RELICT STRUCTUF
AFFECT THEIR EN(

Saprolitic soils, by definition, i
rock. These may include relict f
to repeated movement over ae
slickensides that, in the sapro
low shear stresses. One such :

gradually over time-spans of m
always the case.

Lava flows on Hawaii islan
and chemical weathering withi
ets of grass and small shrubs. 1
asphalt road and the painted wl
already supporting vegetation. *
soil during the South African W
in lateritic material 90 years lat
mudrocks on exposure to the at
over a time period of a few yea
monuments such as the 5000-y
ducts in Europe (see Plate C22
weathering after more than 200

1.13 DETAILED EXAMIN/
PROFILE RESIDUAL

This section will deal in detail w
weathered residual soil profile.
to give examples of the compre
typical residual soil, later in the

Andesite is part of a family
minerals are hypersthenes, augi
andesite in terms of the ratio o
anorthite (Calcium Aluminium
as well as its relationship to otl
andesite considered in this pap
Ventersdorp Supergroup of rocl

The igneous rocks of the
lavas with interbedded zones of
tuffs. The principal minerals of
laths with occasional phenocry
some chlorite, while the volcar
compacted lava fragments.

Alkali

0

0

A

Figure 1.12 Mineralogical compositi
(Capitals denote extrus

The constituent minerals
well-defined colour sequence i
an ancient residual andesite p
altered to chlorite, which imp
the residual soil. As the chlor
structure has oxidized and hyd
soil a yellow or yellowish brov
the soil may be seasonally desic
a characteristic red or reddish
colours do not always all app
and the boundaries between c
for this colour sequence to de
reference to engineering and w

Iron compounds often bec
profile to form ferricrete. As d
itation and gradual accumulat
soil mass. Its development dep
annual fluctuations in water ta
able, compact, brownish accu
staining in the soil. Figure 1.1
the Johannesburg area of Sout
hillwash or colluvium that oft
hardpan or ferricrete is presen

The statistics of the geotec
trends, but this is because the w
of the test holes used for comp

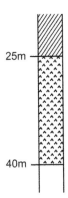

Figure 1.13

andesite soils can vary from cla~
Index above 50) to silty soils (cl:
10). Void ratios can be high (gre
likewise can vary from C_c of o
properties will be given in later

1.14 AN INTRODUCTION
OF UNSATURATED

Because residual soils are foun
rainfall, as well as in semi-arid a
ics of unsaturated soils follows
explanation of unsaturated soil
unfamiliar with the terms and c
The state of unsaturation in
of the stress in its pore water (
having a degree of saturation c
of the stress in its pore water, i
water pressure of less than $-\gamma_w$
in contact with water at a pressu

a lake or pond, exceeds the air
of the year. Such areas cover a
the unshaded and desert areas

1.14.1 The effective str

The pores of most unsaturate
The water exists in capillary le
air-water interfaces are curved
ensure that the pore air pressu
tensions in the air-water meni
soil, as may any salt dissolved

The effective stress equati
form:

$$\sigma' = (\sigma - u_a) + \chi(u_a - u_w)$$

in which u_a and u_w are, respect
ical parameter representing th
that contributes to the effecti
et al., 1960).

The characteristics of this
rium of a typical interparticle c
packed spherical particles (Figi
effective stress equation applie
air enters the pore space, the c
at an interparticle contact car
rium at the contact point, the
P divided by $4r^2$) will be:

$$\sigma_g = \frac{\pi A}{2} \left\{ \frac{pA}{2} - \frac{T}{r} \right\}$$

where $A = (\tan\theta - \sec\theta + 1)$ ar
air becomes an occluded bubb
$\sigma_g = \sigma' = \chi p$ and:

$$\chi = \frac{\pi A}{2} \left\{ \frac{A}{2} - \frac{T}{rp} \right\}$$

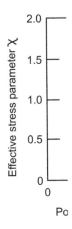

Figure 1.14 (a) Forces at an interpa
of effective stress param
unsaturated state, (c) C
unsaturated soil.

Figure 1.14b shows the vari
which it will be noted that based
rises by a factor of 1.57 (or $\pi/2$

Figure 1.14c shows some va
silt which support the general fo

In most, but not all cases,
atmospheric (i.e., zero gauge) a
will be in a state of absolute tei
strengths of pure water of up to
it is relatively simple to show t
sustain tensions of about 250 kl

saturation of the soil is close to

may be developed in unsatura

familiar with the iron-hard co

desiccated clays of up to 2.8 M

Another advantage of uns

partly filled with air, changes

pore fluid. Thus the shear strei

application of a surcharge stre

An unsaturated soil for wh

soils, viz:

$$\sigma' = \sigma - u_w$$

The parameter χ is obviously i

with S. A number of empirical

the most useful appears to be

et al., 2004). This predicts χ a:

suction at which air would en

entry suction. Algebraically, th

$$\chi = (s/s_e)^{-0.55} \quad \text{if } s \geq s$$
$$\chi = 1 \qquad\qquad \text{if } s < s$$

In Figure 1.15, the bold s

points were measured in tests

1.14.2 Soil suction

In many situations involving u

zero (gauge). As mentioned ab

$-u_w$ (often written as s or p'')

the soil water has two compor

The solute (or osmotic) suc

This arises from the presence

osmotic or solute pressure. Tl

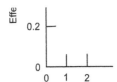

Figure 1.15 Effective stress param

which a pool of pure water mus
semi-permeable membrane (i.e.,
a pool containing a solution ide

The matrix suction

This is the negative pressure to
the soil water must be subjecte
through a permeable membrane
molecules). The matrix suction
apparatus where the water in t
the soil water by a water saturat
American English, the word "m

The total suction

The presence of a solute in wa
librium with the solution whic
to the concentration of the solt
reduction of the relative humidi
the solute suction.

The existence of a tension
humidity in equilibrium with it
of the relative humidity of the a

θ = absolute temperature, K

m_w = molecular mass of water

h = the height of a suspende

would be in equilibrium with

At a standard temperature of 2

$$u_w = 311 \log_{10}(RH) \quad [MP$$

in which (RH) is expressed as

According to Raoult's law

of water containing a dissolve

$$RH = \frac{m_w}{m_w + m_s}$$

m_w is the number of moles of v

solute). For example, suppose

The molecular mass of water is

Because NaCl dissociates into

$$RH = \frac{55.56}{57.56} = 0.965$$

From equation 1.9a:

$$-u_w = 311 \log_{10}(0.965) =$$

This will be the solute suction

solute suctions.

The relative contributions

remain to be resolved. Richard

written as:

$$\sigma' = \sigma + \chi_s p_s'' + \chi_m p_m''$$

in which p_s'' is the solute suctic

p_m'' the matrix suction

…solu.. ….. operates ……

pores of the soil, thus causing €
in this way affects the matrix su

Recent experimental work
that solute suction has no direct
it indirectly by attracting or ret:
suction, and in turn controls the

A practical example of this
Namibia. The roads in the town
irrigated using sea water and th
irrigated with sea water. Altho
Benguela current ensures that, €
fog particles are absorbed by th
and engendering sufficient matr:
result is excellent low-traffic su
Plate C23 shows one of these ro
ening visible in the wheel paths
giving the roads the appearance

1.14.3 The suction-wate

The suction water content cur
unsaturated soils what the cons
the e-σ' curve is the relationship
and consolidation pressure, σ',
or s) and water content w. In a
is the effective consolidation p
vertically. In an unsaturated so
plotted horizontally, with the su

Figure 1.16a is an example
a SWCC. It applies to an undis
represents the virgin compressi
compression line. The reductio
area ABC represents the irrever:
represents 3.2 kJ/m^3 of clay). Th
(0.5 kJ/m^3). Without physically
3.2 kJ/m^3, it is not possible to r

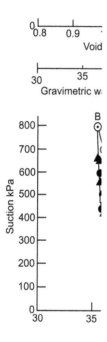

Figure 1.16 Similarities between cor
(a) Hysteresis during co
under mechanical stres
consolidated clay under

Figure 1.16b shows the S
solidated. The similarities be
though the maximum applied
Figure 1.16a (800 kPa versus 8
26 kJ/m^3 of hysteretic energy.
the re-wetting/redrying paths F
guishable hysteresis. In other v
to mechanical consolidation/s
once the clay de-saturates, th
In particular, it is not possibl
hysteretic energy that has beer

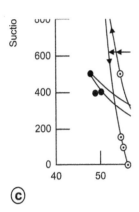

Figure 1.16 (c) Suction/degree of satu
Futai *et al.*, 2004).

Figure 1.16c shows SWCC
et al., 2004). (The profile is s
contained about 45% of clay a
The huge difference in the sucti
differences in particle size distril
terms of degree of saturation S =
1.5, and assuming G = 2.7 for
have been about 55%. For the
uration would have been 35%.
at desaturation would have bee
soil mechanics will be found in

REFERENCES

Aitchison, G., Ingles, O.G. & Woo
utory factor in the failure of ear
Eng., Adelaide, pp. 275–286.
Bell, F.G. & Walker, J.H. (2000)
dispersive soils in Natal, South A

Bright, G.L. (2000) Geoenviron̳
    municipal solid wastes in water
    S.L. & Fredlund, D.G. (eds) ̳
    ASCE, Reston, USA, pp. 36–80

Brink, A.B.A. & Kantey, B.A. (1
    Southern Africa. In: 5th Int. Co

Brink, A.B.A. (1979) Engineering
    South Africa.

Bromhead, E.N., Del Preto, M., R
    problems in Basilicata, Italy, re
    Symp. on Eng. Characteristics (

Casagrande, A. (1965) Role of ca̳
    Mech. & Found. Eng. Div., AS̳

Desai, M.D. (1985) Geotechnical ̳
    H.B. (eds) Sampling and Testin̳

Dudley, J.H. (1970) Review of co
    96 (SM3), 925–947.

du Toit, A.L. (1954) The Geology

Elges, H.F.W.K. (1985) Dispersive

Ernst, W.G. (1969) Earth Materia

Falla, W.J. (1985) On the Signifi̳
    Characteristics of Residual Soi̳
    University, Johannesburg, Soutl

Fitzpatrick, R.Q. & le Roux, J.
    topo-sequence. J. Soil Sci., 28, ̳

Fredlund, D. & Morgenstern, N.R
    Eng. Div., ASCE, 103 (GT5), 4

Fredlund, D. & Rahardjo, H. (19

Futai, M.M., Almeida, M.S.S.
    conditions of a tropical satura
    169–1179.

Gerber, A. & Harmse, H.J.vM. (1
    by chemical testing. Civ. Eng. S

Gidigasu, M.D. & Mate-Korley
    tions. In: 8th African Reg. Co
    pp. 267–273.

Gonzalez de Vallejo, L.I., Jiminez
    of the tropical volcanic soils of

Hight, D.W., Toll, D.G. & Grace,
    In: 2nd Int. Conf. on Geomech

new evidence. *Int. J. Geotech.,*

Knight, K. (1961) *The Collapse*
    Witwatersrand University, Johan

Leong, E.C., Rahardjo, H. & Tang,
    Singapore Residual Soils. In: *Int.*
    Natural Soils. Singapore, Vol. 2,

Li, W.W. & Wong, K.S. (2001) Ge
    Engs., Singapore, 41 (3), 10–20.

Little, A.L. (1969) The engineering
    Mech. & Found. Eng., Mexico C

Lumb, P. (1962) The properties of

Madu, R.M. (1980) The use of cl
    identification. In: *7th African*
    pp. 105–116.

Mason, B. (1949) Oxidation and re

Medina, J. (1989) Tropical soils in
    Eng., Rio de Janeiro, Brazil, Vol.

Mitchell, J.K. (1976) *Fundamentals*

Morin, W.J. & Ayetey, J. (1971) Fo
    Reg. Conf. Soil Mech. & Found.

Netterberg, F. (1994) Engineering g
    Int. Assoc. Eng. Geol., Lisbon, F

Paige-Green, P. (2009) Dispersive
    Course on Problem Soils. Johann

Pings, W.B. (1968) Bacterial leachi
    2 (3).

Reeves, G.M., Sims, I. & Cripps
    Geological Soc. Spec. Pub. No. 2

Richards, B.G. (1966) Significance
    tion to design of structures bui
    Permeability and Capillarity, Atl

Ruxton, G.P. & Berry, L. (1957)
    Hong Kong. *Bulletin, Geol. Soc.*

Schwartz, K. & Yates, J.R.C. (198(
    African Reg. Conf. Soil Mech. &

Sherard, J.L., Dunnigan, L.P., Dec
    dispersive soils. *J. Geotech. Eng.*

Strakhov, V. (1967) *The Principles*

Sueoka, T. (1988) Identification a
    weathering index. In: *2nd Int. Co*

for Soil Mech. & Found. Eng.,

Viljoen, M.J. & Reimold, W.U. (1?
Heritage. Johannesburg, Minte

Wagener. F.v.M. (1985) Dolomite

Weinert, H.H. (1964) *Basic Igne*
218. Pretoria, South Africa, CS

Weinert, H.H. (1974) A climatic
Géotechnique, 24 (4), 475–488

Wesley, L.D. (1973) Some basic er
Indonesia. *Géotechnique,* 23 (4

Wesley, L.D. (2010) *Geotechnical*

Yamanouchi, T. & Haruyama, M
Memoirs, Faculty of Eng., Kyu

Zaruba, O. & Mencl, V. (1976) E

Zhang, G., Whittle, A.J., Nikolir
engineering properties of the ol
terization & Engineering Prope

2.1 MICROSTRUCTURE /
RELATED TO WEATH

The microstructure and minera
modes of soil formation and oc
relative scales:

1 ped and within-ped or intra
2 grains and intergranular vo

Peds are natural aggregatio
scale ranging from centimetres
interaction of the soil grains, an
drying or various methods of m

Physical breakdown and ch
the formation of tropical and res
exposure to water and changing
to the parent material from the l
categorized into decomposition

These processes may procee
on the climatic conditions and th
Both mineralogy and microstru
A wide variety of soils can be p
to keep a broad perspective wh

2.1.1 Decomposition

Decomposition includes the ph
breakdown of constituent mine
products are clay minerals, oxic

Under conditions that inclu
transported soils may be modi
Reaction rates vary so that sor
feldspars in granite) when neighl

2.1.3 Dehydration

Dehydration (either partial or
sesquioxide-rich materials in a
Dehydration also influences tl
total dehydration, strongly cer
be formed.

2.2 MINERALOGY AND
PRODUCTS

Tropical decomposition tends
is by far the most common clay
conditions, halloysites will be
removed to the extent that free
of the silica produced in the soi
present and/or remaining are st
form a mineral depending on
formed in very strong oxidizin;
there are conditions of contin
characteristic of soils from cert
identified in African lateritic s
kaolinite is present. Montmori
eral in soils residual from basa
residual from sedimentary roc

Other mineralogical comp
may be relatively rare and/or of
anatase, mixed-layered kaolini
rillonite may be present in lat
transitory mineral in one part

Tropical weathering of vo
phone, a virtually amorphous
content. Allophane may be ide
of plasticity properties upon d

Clay minerals tend to be (
minerals such as goethite and/

The classic weathering pr...
widely, and been subjected to d
Within the soil layer, there may
The soil profile is typically struct
enriched intermediate horizon, (
to maintain a very disciplined pr
adequately to record residual so

2.3 DETERMINATION OI

Soil mineralogy can be assessed
developed for particular purpos

- X-ray diffraction,
- Thermo-gravimetry,
- Optical microscopy includii
- Scanning or transmission
  spectral element identificati

Mineralogical identificatioi
and procedures. Often, combii
definite identifications. The proi
and measurement process usual
be fully understood to be of gre

The x-ray diffraction (XRD
appropriate for minerals with
using oriented or randomly seli
without resin to fix the samples
containing significant quantities
by ethylene glycol) causes chang
of montmorillonites.

Thermogravimetry (TG) idi
dration takes place through a r
for certain simple minerals wh
signatures.

Optical microscopy (OM),
a well established assessment pi

of thin sections.

SEM imaging can reveal n
ticularly useful tool for micro
microprobe, for example energ
made of elemental constituent:
abundances of elements, it is t
material at the point. For man
forms that are common in resi
tification unless this process is
which reveals crystal form.

TEM imaging can reveal :
used to refine knowledge of th
used to make chemical compo
visualize the layering within cl:
take place during the weatheri

2.4 MICROSTRUCTURE

Soil structure, fabric, and text
grains and peds. These arrange
behaviour. In conventional so
nized for a long time, and has l
fabric relevant to engineering l
without magnification).

Microstructure embraces n
tigations of soil microstructure
The microstructure of residual
ments, which can lead to a b
performance of such soils.

Collins (1985) extended m
els presented in Figures 2.1 anc
at three levels:

- elementary level (Figure 2.
- assemblage level (Figure 2
- composite level (Figure 2.:

Assemblage level

Individual particle assemblages (ipa) formed b̶

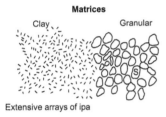

Matrices

Clay   Granular

Extensive arrays of ipa

Figure 2.1 Elemen

Composite microfabric

matrices
+
aggregations → assemblage
+   network
connectors

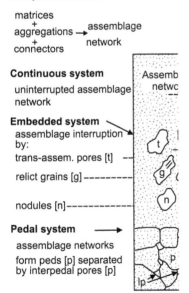

Continuous system

uninterrupted assemblage
network

Embedded system

assemblage interruption
by:

trans-assem. pores [t]

relict grains [g]

nodules [n]

Pedal system

assemblage networks

form peds [p] separated
by interpedal pores [p]

Assemb
netwo

Figure 2.2 Compos

of particular soil microfabrics
concepts are important and sl
order to understand how the
fluid flow is likely to occur thr

2.5 MINERALOGY AND
GEOTECHNICAL PR

Mineralogical influences on en
ple, soils which contain halloys
to drying and also to manipul
presence of this clay mineral m
more difficult than usual.

Effects of drying

A distinctive feature of halloy
the irreversible change in soil
ral moisture content (see also
mineralogical changes caused
testing of known or suspected
be performed with the intende
in its natural state, as it usually
be applied to soil that is as clo

Effects of reworking

Similar experiences can be ob
which have sustained subseque
composed of volcanic clasts ma
but as a high plasticity clay v
reworking (e.g., as in cut-to-fil

Reactivity to stabilizing ag

Soils with kaolinite as the clay
to medium plasticity and pern
present (or even halloysite) the

Effect of secondary minerals

The influence of sesquioxides is
cement the particles, they tend
is only very weak cementation,
interactions between clay partic
presence of goethite/haematite a
by the process of stabilization, f
ing the strength of the soil. Ki
stiffness in a saprolite to the br
erals and/or sesquioxides. This i
the unsaturated soil.

Baynes and Dearman (1978
of weathered granites were due t
observed in the microstructure o
similar features in weathered gr
that a highly-cemented microst
reactive to the process of lime :
form of amorphous aluminous :

Dispersive behaviour

Dispersive soil behaviour (see se
subjected to cyclic wetting and
sodium-dominated. Simmons (1
was the most identifiable factor
erosion. These clay minerals co
ration zone of seasonal ground
rather than clay chemistry, had
found to be greatest where silt-s

In summary, residual soils r
and mineralogy. Where specific
fied, corresponding influences c
important principle is that resid
istics than index tests (on the re
between index tests and the eng
Soil properties must always be a

cnange as a result or weather
that depends on the minerals
magma have the highest inter
most unstable and will break
the last minerals to crystallize
the most resistant to weatheri

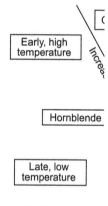

Early, high
temperature

Hornblende

Late, low
temperature

(a) The order of c

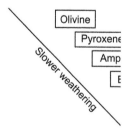

Olivine

Pyroxene

Amp

E

(b) The order of w

Figure 2.3 Crystalliza

... not form part of the origin...
secondary deposition of quartz
find quartz veins in residual an
regular progression of minerals
of weathering to those charactei
composed of a number of succes
different mineralogy and each o
before being covered by the nex

The second example is sho
weathered gneiss in Brazil. Here
quartz and kaolinite with a rela
minerals that feature prominent

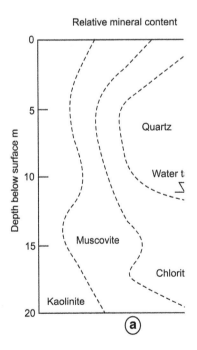

Relative mineral content

Depth below surface m

Quartz

Water t

Muscovite

Chlorit

Kaolinite

(a)

Figure 2.4 (a) Mineral distribution in
profile of weathered gneis;

... for long periods.

Apart from the relatively
residual soils, drying also affec
characteristics of residual soils

2.7.1 Water content

The conventional test for the
of water when a soil is dried
110°C. In many residual soils,
structure of the minerals prese
this water may be removed by
is driven off. This is illustrate
water content of different dryi
105°C. As shown, the water co
temperature. Figure 2.6 shows
progressively for four residual
to 40°C (and relative humid
105°C. The effect is even more
(Terzaghi, 1958; Frost, 1976; '
atic as this may take an extren
procedure is therefore recomm
content determinations. One s
weighings show no further los
in the normal way. The secon
at a temperature of no more
30% until successive weighing
results should then be compa
tent obtained by oven-drying
and is driven off at high temp
should therefore be excluded f
detected using the two differen
tent determination (including
be carried out by drying at the l
at 50°C and 30% relative hum
should be used.

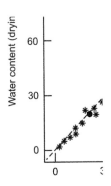

Figure 2.5 Effect of d

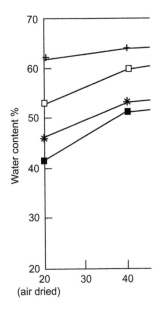

Figure 2.6 Effect of drying temp

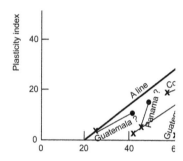

Figure 2.7 Effect of drying on p

2.7.2 Atterberg limits

In addition to the problem of w
problems may be experienced v
optimum water content and ma
of pre-test drying, and of dura

The effect of pre-test dryin

The effect of air-drying specim
than testing, starting at natura
ent decrease in the liquid lim
1986; Wesley & Matuschka, 1
Figure 2.7.

According to Townsend
attributed to:

1   increased cementation due
2   dehydration of allophane
3   both 1 and 2 above.

In order to be meaningful
be performed without any for
of drying is unavoidable, for

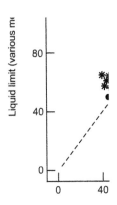

Figure 2.8 Effect of m

this should be noted on the lab
duration of drying.

Usually, Atterberg limit tes
425 μm. The above recommen
far as possible, all large soil par
with the fingers through a 2.38
without breaking them up.

The effect of duration and n

In general, the longer the durat
the soil prior to testing), the la
the larger the plasticity index. T
break down of cemented bonds
formation of greater proportion
An extreme example of the effe
by Figure 2.9 which shows the i
Liquid Limit cup from 20 after
The soil in this case was a resid

In order to address this pro

Five test specimens should l
suitable for liquid and plastic lin

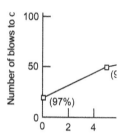

Figure 2.9 **Effect of time of miᐳ**

should be used, and preferablɣ
climates and with most soils t
mixing time should be standa
be kept in sealed containers fo
water content to equilibrate w

The liquid limit should bɛ
Standard 1377, Part 2) with a
A sub-sample from each of the
ing the water content, using th
described in section 2.7.1 abov
continuously for a further 25 ɾ
nificant difference (i.e., >5% ɕ
for 5 minutes) between the liqɩ
indicates a disaggregation of tʰ
confirmed by a repeat of the aʰ

1   Limit the mixing times to
2   Make use of fresh soil for ɛ
    as in compaction tests.

The soil should be broke
grinding. The soil should be iɾ
through a 425 μm sieve until tʰ
be collected and used for the ⱥ
should be dried and weighed.
should be calculated and recoɾ

(his ....), ...
alent "soil water" can be prepa
water and soil (about a 3:1 by v
to settle, the clear water is decan

2.7.3 Particle relative u

The particle relative unit weight
volumetric parameters such as
unusually high or unusually lo
voidless solid particle of a heavy
of elementary particles. It is th
using an accepted standard test
dry density, porosity, etc.

The soil to be used in this te
of the soil should be avoided as
measurements at natural water
be calculated by drying the soil
completed. Depending on the
above, it may be necessary to ai

2.7.4 Particle size distri

As with the above parameters,
affected by certain aspects of sa

1 **Effect of drying:** The most w
   age that is reported as the c
   to testing should be avoided
   one for determining the wa
   the other for the psd test.
   of dispersant such as dilute
   washed through the standa
   been added, the washing sh
2 **Chemical pre-treatment:** Th
   with hydrogen peroxide is

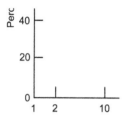

Figure 2.10 Grain size analysis of a
(full line).

considered necessary to el:
with hydrochloric acid is u

3   **Sedimentation:** It is essenti
to carrying out a sedimen
phate is suggested. In som
may be required. If the a
trisodium phosphate shou
freshly made before use in

Because of the tendency of
(IPAs) or peds (see Figure 2.1)
analysis. This is illustrated by
example of particle size analy
size distribution is that of a sili
that of a sandy clay with an ex

2.8   CLASSIFICATION O

There are specific characterist
methods of soil classification
Unified Soil Classification Syst

1   The clay mineralogy of so
compatible with those nori
according to existing syste

by the translational definition
converts rock into soil there will
shown in Figure 1.9, the upper
lower zone will behave as soft
applied only to the upper two zc
soil with clear structural feature
used for features present in the
upper zone have usually been o
other agents of disturbance.

2.8.1 The place of the U

The USCS divides soils, primaril
silts (M) and clays (C) and org
because of the origin of residua
"silt" cannot be applied to resid
use the Casagrande plasticity ch
between "soil liquidity" (via the
index $PI = (LL - PL)$, where PI
L, H and I denote low, high an
be used to display the same rela
in Figure 2.11b (after Wesley, 1
clay mineral on the position of
a given residual soil can be use
likely predominant clay mineral
2.11b from the original Casagr
accommodate the very high liq
soils of volcanic origin.)

 The USCS originally appea
projects" (Casagrande, 1947; Ta
on the California Bearing Ratio
the soil classification. (Before p
sively by the US Airforce for de
the Second World War. Consid
transported soils, it would be in
war-time application.)

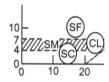

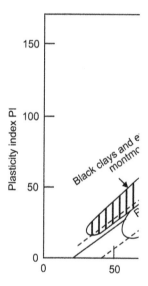

Figure 2.11 (a) The Casagrande plas
on plasticity chart.

2.8.2 Wesley's classific

Wesley (1988) proposed a pra
mineralogical composition and
system is intended to provide a

Table 2.1 Characteristics of residual soil groups (After Wesley, 2010).

Group				Gr
Major Group	Subgroup	Examples	Means of Identification	Pr
GROUP A Soils without a strong mineralogical influence	(a) Strong macrostructure influence	Highly weathered from acidic or intermediate igneous or sedimentary rocks	Visual inspection	Lä "s
	(b) Strong microstructure influence	Completely weathered from igneous or sedimentary rocks	Visual inspection, evaluation of sensitivity	b) Eä ar pi in

2.9 EXAMPLE OF CLAS

As an example, the soil formii
Figure 1.13 will be classified. ˙

1   The clay minerals compris
    and chlorite, together with
    with depth, but cannot be
    soil. Thus the soil falls wit
2   The saprolitic joints in th
    of the soil, so it falls into
    resulted mainly from cooli

Hence the soils classifies as: A

REFERENCES

Arnold M. (1984) The genesis, m
    on Expansive Soils, Adelaide, A
Baynes, F.J. & Dearman, W.R. (1!
    Int. Assoc. Eng. Geol., 18, 91–
Blight, G.E. (1996) Properties of
    Soils, Kuala Lumpur, Malaysia.
Bowen, N.L. (1928) The Evolutic
Brand, E.W. (1988) Evolution of ;
    in Tropical Soils. Rotterdam, A
Brummer, R.K. (1980) The Engi
    Profiles. MSc(Eng) dissertation
Casagrande, A. (1947) Classificat
    June, 295–310.
Collins, K. (1985) Towards charac
    in Tropical Lateritic and Sapro
Falla, W.J. (1985) On the Signific
    Characteristics of Residual Soi
    University, Johannesburg, Soutl
Frost, R.J. (1976) Importance of ;
    Asian Conf. Soil Eng., Bangkol

alumina on selected properties o
Vol. 2, pp. 559–567.

Queiroz de Carvalho, J.B. (1991) M
can Conf. Soil Mech. & Found. I

Richards, B.G. (1985) Residual sc
Sampling and Testing of Residua

Rodriguez, T.T. (2005). Colluviu
COPPE/UFRJ, Brazil. Quoted ii
and Colluvium in Southern Bra
Netherlands, Amsterdam, Elsevic

Rouse, W.C., Reading, A.J. & Wals
Indies. Eng. Geol. 23, 1–28.

Rowe, P.W. (1974) The importanc
Géotechnique, 24 (3), 265–310.

Simmons, J.V. (1989) Preliminary
Irrigation Area using the Scannin,
and Systems Engineering, James (

Taylor, D.W. (1948) Fundamentals

Terzaghi, K. (1958) Design and per
369–394.

Townsend, F.C. (1985) Geotechnic
111 (1), 77–94.

Wesley, L.D. (1973) Some basic eng
Indonesia. Géotechnique, 2 (4), <

Wesley, L.D. (1988) Engineering
Tropical Soils, Singapore. Vol. 1.

Wesley, L.D. (2010) Geotechnical I

Wesley, L.D. and Matuschka, T. (1S
Conf. Geomech. Tropical Soils, S

Sites identified for construction
qualified and experienced profe
sional engineering geologists w
decision on the suitability and
conditions at the site have been

The following information :
engineering development.

3.1 SOIL ENGINEERING

Every civil engineering project sl
in order to establish the differe
boundaries within which they o

- Typical soil profiles, depths
  preliminary estimates of the
  rock and their *in situ* shear
- Any indications of the prese
  which were not revealed on :
- All potential problems with
  expansive or collapsing soil
- Geophysical methods are a
  means of extending the bas
  karst areas gravimetric surv
  sinkhole or subsidence are;
  contrasts. Microseismic me
  Resistivity surveys may be i

3.2 SOIL ENGINEERING

Soil engineering information is n
1 metre) boreholes which can b
hole auger. Holes dug by a b;
considerations limit the depth o

ill with methane and carbon (

or other gases in the case of a r

top of a hole may also cause it

of which are heavier than air.

paper down the hole to obser

Stand well clear of the hole an

a shower of dust and soil part

Safety demands that no-c

observer wear a hard hat and

around his body by means of

If large diameter drilled

may be necessary to explore tl

150 mm) boreholes. The actu

on the characteristics of a par

engineer or engineering geolo

geology and from features visi

In addition to the geolog

during the course of the site su

- any evidence of local seism
  area (available from the lo
- mean and extreme rainfall
  monthly pan evaporation
- maximum rainfall intensit
  years and 24 hours in 100
- mean monthly wind direct
  the maximum one minute
  is preferable, if available).

3.3 DETAILED INFORM

The following detailed inform

ment or foundation excavatio

seepage cut-offs, load-bearing

- The permeability of the f
  soil: The permeability is I

are described in sections 7.

or consolidated drained tria
taken from the field may b
parameters of the soil in te
section 7.2. Whichever metl
lish the shear strength prop
In situ test methods are m
access to the soil from the su
than 5 metres) and in dry c
building is likely to have a
soil, (i.e., wetting of the soi
undisturbed specimens take
effect of changing the groui

- In many cases it may be n
conditions for a developmei
being planned, the characte
long term and short term sta
information not only on the
stresses but also to measure
by means of triaxial shear c

- The compressibility of the s
e.g., loose, potentially coll
be necessary to estimate th
undergo. If compressibility
compressibility or expansib
measured by means of labo

3.4 PROFILE DESCRIPTI

Because all soil sampling involv
tinguish facts related to the soil
inspection or sampling. The eng
state may be difficult to assess,
excavation equipment, drills, o
advance of a drill, or rate of pei

variety of organizations, and
practice. Most of these proced
acceptable recording procedui
1973; Brink, 1979; Cook & N
extended for other purposes. "
rather than for specific purpos
for reasons not earlier anticipa
profile may be used for a diffe
different one is later embarkec

The following procedure I
purposes. It may be supplemei
priate sections of locally appli
choice for use of any such lists

3.4.2 Site records

The following should be recor

- general description of sit
  across site, access routes (i
- dates of site investigation,
- weather during investigati
- precise location details (cc
- all field activities (diary, lc

  Descriptions should be rec

- Moisture condition (M),
- Colour (C) (Best describ
  known Munsell soil colou
  Multicolours, e.g., mottlin
- Consistency (C),
- Soil (e.g., clay, silt, sand, g
- Structure or fabric (zonii
  nodules etc.) (S),
- Origin (transported or res
  blown, delta deposit, fluvi

....

A convenient procedure for
measuring tape at the surface or
orientate himself as he observes t
layers, even when he is unsight
very necessary, as it is difficult t

As the observer descends th
moisture condition, colour, cor
of the soil profile (MCCSSO) sl
thickness of each.

The observer can either not
recorder, or (a better, safer pract
surface who records the observat
under each of the categories M(
and act if the observer encount
or a partial or complete loss of c
poisonous or asphyxiating gas.

(<u>M</u>CCSSO) moisture conditi

This is recorded as dry, slightly n
provides a useful indication of
used in construction as fill. Dr
water to attain the optimum m
near the optimum moisture co
soils are generally only found b

(M<u>C</u>CSSO) colour

Colour is used for describing the
in different holes. Colours are
the same colour in his mind's eye
a standardized soil colour chart
small in size and suitable for carr
chart which has the segments of
The actual size of the pocket ch

The natural colour as seen i
should be noted, e.g., "light grey

gravelly soils.

(MCC__SS__O) structure

Intact indicates an absence of
    Fissured indicates the pres
stained with iron and mangan
    Slickensided indicates the
and sometimes striated. To o
carefully break it apart to re
Plate C20).

Table 3.1 Categories of soil consiste
    .

Granular Soils (usually clay free)

Very loose	Very easily penetrated l geological pick
Loose	Small resistance to penetration by geological pick
Medium dense	Considerable resistanc penetration by geologi pick
Dense	Very high resistance to penetration by geological pick
Very dense	High resistance to repeated blows of geological pick
Extremely dense	Repeated blows of geological pick make no impression

and range of sizes should be rec
if present. The matrix should al
described as this often aids the i

> *well-rounded* (nearly spheri
> *rounded* (tending to spheric
> *sub-rounded* (corners roun
> *sub-angular* (corners remov
> *angular* (corners sharp or ir

Gravel consists of particles
of description follows that for t
Sand consists of particles l
Most of the particles sizes are v
Silt consists of particles wh
Individual particles cannot be
gritty when rubbed against the
Clay consists of particles sn
soapy when wet.
Most natural soils have a c
describing such a soil, the adje
the predominant size, e.g., a si
approximately equal proportio
some clay.

(MCCSS<u>O</u>) origin

An attempt should be made to
soil profile. On many sites a pel
between transported and residi
drainage which, if drainage is
providing a free flow of water. I
be predicted from experience w

Transported soils

Possible origins of transported
highest to lowest topographic e

Alluvium or gulley wash	flow in: river stream gulley
Lacustrine deposit	stream (pan (por pond lake
Estuarine deposit	rivers ar
Aeolian deposit	wind
Littoral (beach) deposit	waves

*If any of these deposits are residual soils (see Plate C14).

Residual soils

A knowledge of the local geol
to the origin of residual soils
preservation of the primary (o
parent rock, e.g., bedding plan

It is sometimes possible to
the mineralogical compositior
granite will contain quartz gra
position of feldspars. Those fc
will contain no quartz (other t
entirely of clay and silt. Amyg
volcanic lava such as andesite
parent rocks are usually easy
bedding structure, particle size

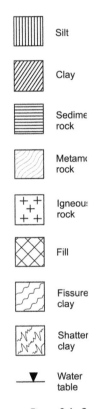

Silt

Clay

Sedime
rock

Metam(
rock

Igneou:
rock

Fill

Fissure
clay

Shatter
clay

Water
table

Figure 3.1 S)

3.4.3   Recording of soil p

The left hand side of the profi
drawn to scale, which shows th
are indicated by the standard s
record the presence of such feat
the full depth of penetration of

To the right of each stratur
be recorded, as shown in the ex

0-0.1m   Made ground

1.3m   Moist, yellowish-brown, loose silty sand - Hillwash.

Moist, dark red, firm to stiff, partly lateralized silty clay. Residual diabase.

2.7m   Moist, yellowish-orange blotched dark red, stiff to very stiff, intact, talcose clayey silt with lateritic cementing. Residual diabase. Partly lateralized to 6.6m.

0m

1.0m

1.5m

2.5m

0m –

0.6m –

Very moist, black, very soft, slickensided, clay-silt, marsh soil

Water table

Wet, dark brown, soft, slickensided, clay-silt; marsh soil.

Poorly developed pebble marker

over the bottom half metre and
geofabric. If possible, the hole
clean sand or gravel before bac
spoil that came out when diggir

It may take several days or
stabilize. The water level in the
trical dip meter, or if not availab
the standpipe to the bottom. If
be heard when the tube is blow
the tube, the level of the water
sound stops. The water table de
the standpipe, measured to grou

3.5 SIMPLE *IN SITU* TEST

It is usually convenient to augm
in section 3.4 by simple *in situ* t
time as the visual assessment. Tl
are particularly useful for this p

Sampling can also be carri
testing. These are usually distu
the purpose of the tests from t
quantity required will vary from
size analyses to 20 to 30 kilogra
and the soil will be remoulded
tests, the samples are usually co
carefully labeled and the bags
thin rope. Plastic bags (plastic
preferable as the soil should be
until it is tested. All Atterberg I
in situ water content, as air-dryi
its Atterberg limit and compact

It may also be convenient to
to establish shear strength and c
course, be taken above the wate
trim a pedestal of soil out of the
space) to fit an open-ended cyli

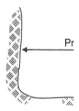

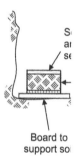

S
ar
se

Board to
support so

Figure 3.3 **Procedure for cutting ar**
test hole.

300 mm long piece of rigid pl
over the soil pedestal by trimm
pedestal, the pedestal is cut of
the cylinder can then be sealed
wax, or even by wrapping in s

3.6 TAKING UNDISTUR
LABORATORY TEST

The purpose of taking "undis
to the stress changes and dra
field by the prototype structur

In situ
_____

Figure 3.4 Changes in stres

Therefore it must represent the *a*
and water content. Sampling pr(

3.6.1 Sampling of satura

The relationship between void r
for a saturated soil is:

$$e = wG$$

Hence a saturated soil is incon
does not change. In a clayey so
the surface should not result in
water by capillarity. The soil wi
pressure as a result of the releas
by Figure 3.4.

3.6.2 Sampling firm to s

The usual method is to use a 76
like that illustrated in Figure 3.
sampler into the soil at the bot
to "overdrive" the sampler, i.e.
further than the length of empt
off the sample at its base. The sa
into a plastic film tube, placed
tube and waxed to seal and sup
before testing. The soil must be
as it moved into the tube, i.e., fi

Great care must be taken
laboratory. Sample tubes shoulc
and be packed in a shock-absor
or failing these, sawdust or woc

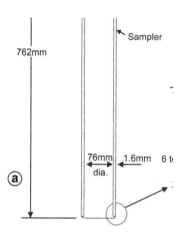

762mm

Sampler

76mm
dia.
1.6mm   6 t

(a)

Figure 3.5 (a) Thin-walled tube opei
table. (b) Thin-walled sta
water table.

3.6.3 Sampling soft sat

When an open-drive sampler :
side restriction provided by the
stages of penetration. Howeve
inside the tube and the sample
by the progression A, B, C. T
length, piston samplers, as sh
piston is initially flush with th
When the sampler is ready for
sampled. The piston is then cla
it. Thus, while the sampling tu
at the original level of the top
cannot change during samplir
is incorporated in the sample

Figure 3.6 Effects of pushing

empty sampler tube by hydrosta
the sampler is being pulled up :
procedure is known as fixed pis

Another system sometimes
is pushed forward but the rod:
when the sampling tube is pusł
ple. When the sampler is withdɪ
retained in the tube.

Although the piston sampl
been used successfully to sampl

3.6.4   Important soil san

Open-drive and piston sample
quantified and are of importanɑ
The amount of disturbance cau
large degree on the "area ratio"
of the sampling tube and the cɪ
the sampling tube shown in Fig

$$C_a = \frac{(D_w^2 - D_e^2)}{D_e^2} \leq 10\%$$

It has been shown that, for laɪ
soil by the tube itself is very sn
1948). To achieve such a low arɑ
seamless, high strength steel tuɩ

Another cause of disturbanɑ
the length of sample which can ɭ
friction can be reduced by provi

Figure 3.7 Dimensi<

having a low coefficient of fri
D_e slightly smaller than the in:
expressed by the inside clearar

$$C_i = \frac{(D_i - D_e)}{D_e} = 0.5\% \text{ to}$$

The Recovery Ratio RR is defi

$$RR = \frac{\text{Length of san}}{\text{Distance sampler w}}$$

The Recovery Ratio should li
ples with RR's outside the lim
compressibility tests.

3.6.5 Sampling very sti

In order to sample very stiff an
a satisfactory undisturbed cor
barrels, Figure 3.8 shows:

(a) A single tube core barre
hard metal studded bit th
barrel. Rotation of the b
into the rock. Drilling fl
rods both cools the bit ar
hole. The rock core mus
drilling fluid, flowing ov
(b) Shows a more refined dot
and protects the core aga

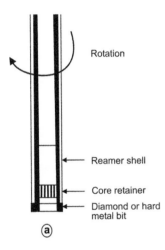

Rotation

Reamer shell

Core retainer

Diamond or hard
metal bit

ⓐ

Figure 3.8 Core barrels for rotary dr
sampling but unsuitable fo
core barrel, suitable for so
soft rocks and hard soil.

(c) The triple core barrel is
rocks. The third tube is a
tionary. The second, or in
ahead of the cutting bit. T
core, as the bit cuts into the

When drilling cores from s
water, combined with stress rel
a drilling fluid, instead of water

REFERENCES

Brink, A.B.A. (1979) *Engineering* (
Building Publications.
Burland, J.B. (1958) *A Simple S*
project, Department of Civil En
South Africa.

4.1 THE COMPACTION P

Compaction is a process whereb
compaction results from the wc
the specific purpose of compact
e.g., to dump more material. *Rc*
of a surface by a vehicle specifie
compactor or roller. The energy
moving down as it compresses
e.g., resulting from the repeatec
gyration of an eccentric weight a
is a fourth method of compacti
four ways of expending compac
surfaces that contact the soil in
tyred rollers as well as grid-roll
technology of compaction and t
compaction of residual soils.

Compaction occurs becaus
thus expelling air and reducing
to result in the expulsion of wa
during compaction and the resu

For a given energy input anc
on the water content of the fill
which a given energy input wi
method of input is changed, bo
density will change. As shown
maximum dry density increases
decreases.

The largest effect on dry c
Thereafter as indicated by Figu
worthwhile applying more than
obtained with this number of pa
too low, the energy input per ro
for the soil type (see also sectior

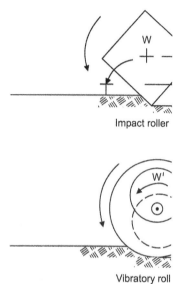

Impact roller

Vibratory roll

Figure 4.1 Four ways of expending c
h, expending energy Wh c
W'. (d) Pounding soil sur

Compaction results from
stresses on the fill material. Th
reduce or disperse with increasi
that occurs is illustrated by Fi
ciable effect at a depth greater

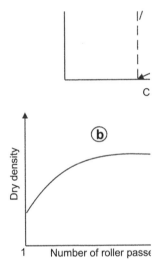

Figure 4.2 Relationships between: (a)
(b) roller passes and dry d

reasonably uniform state of com
be spread in layers not exceedii
a thickness of about 200 mm. T
and pounding (Figure 4.1c) are
(but is not always) attained dov

Smooth-wheeled and pneu
earthworks such as earth dam
compaction required by road la
deep loose sandy soils, and mi;
road or railway embankment, i
rollers are useful if the soil cont
partly-weathered residual soil ai
by compaction. Footed rollers w
layer being compacted are favc
layers for the construction of ir
1989).

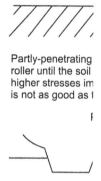

Partly-penetrating
roller until the soil
higher stresses im
is not as good as t

F

Figure 4.3 Four kii

Because smooth wheeled r
used to finish off the top surfa
will be covered by a geomemb
promotes the close contact be
functioning of a composite soi

Laboratory compaction te
optimum water content and t
paction energies. Laboratory v
for roller compaction as the me
ally by repeated blows of a dr
small rigid-walled mould) diff
because design options usuall)
also important to study labora

4.2 CONSEQUENCES O

It can be very difficult and exp
of the time that may be invol\

camber in a road surface, a
for rectification.

4.3 THE MECHANISMS O

Essentially, compaction is a pr
soil, reducing the air-filled void
closer together. It is the process (
increases the dry density and he
ibility and also reduces the per
these effects, both shear and co
that is what requires the expenc

At low water contents the
is relatively high. The air-filled
the soil. With a given expenditu
low compacted dry density can
resistance of the soil to compact
air-filled voids decrease, the resi
voids become occluded, or seale
pore space, This point corresp
and optimum water content for
optimum point onwards, as w;
in the voids, while the air conte
density of the soil decreases pro
shows the "zero air voids" line,
would contain no air.

A set of relationships betwe
illustrated in Figure 4.4. It must
compaction. The relationships ;
using standard Proctor compac
shales at the site of the Mangla
tent versus dry density γ_d, shear
degree of saturation S for each
water content increases, S incr
mum water content, and then

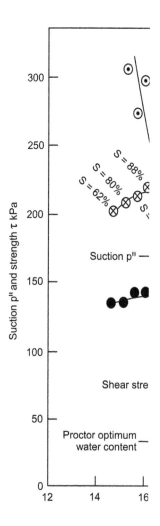

Figure 4.4 Relationships between co
τ for a clay residual from

...compacted in 3 layers with 27...

7.5 J × 27 blows × 3 layers × 10...
hammer is a standard, most soi...
compaction machines that distri...
over the circular surface of the...
hammer is now relegated to the...

To establish a dry density v...
loose soil are prepared at a seri...
compacted into a mould with th...
the soil surface flush and level v...
removed from the mould and sp...

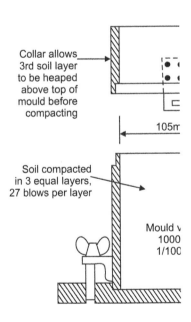

Collar allows
3rd soil layer
to be heaped
above top of
mould before
compacting

105m

Soil compacted
in 3 equal layers,
27 blows per layer

Mould v
1000
1/100

Figure 4.5 The 105 mm diameter sta
compaction hammer.

Most guides to compaction s[
be stored in sealed container:
throughout the specimen befoi
moisture homogenization proc
should be allowed. The water
and soil aggregations, but also
Even after the 26 days for ho
knowing if the properties of tl
layer spread at its *in situ* watei

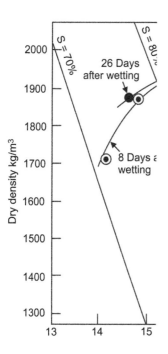

Figure 4.6 Time taken to achieve ;
clayey soil.

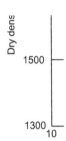

Figure 4.7 Eff

4.5.2 Soil aggregations (

Hard soil aggregations or clods
larger and more clods there are
paction curve will be masked.
larger (>20 mm) initial clods res
mum and require a larger opti
(<5 mm) shows a higher dry de
lower optimum water content.
compaction, an even lower opti

4.5.3 Other treatments
compaction curve

Figure 4.8 (Gidigasu and Dogbe
from site at a water content of
allowing 7 days for disseminat
(the "natural state", A in Figure

The same soil was air dried,
the "natural" sample, and the r

When, however, sub-sampl
a very different compaction cur
pulverizes under compaction, it
from particle break-down cause
and then rewetting and compa

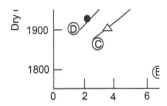

Figure 4.8 Effects of various trea

probably because the oven dryi
or metallic hydroxides contain

It is therefore essential to
between taking the sample in t
compaction testing. If the sur
surface layer should be discar
stored in sealed plastic bucket
curve. Use a fresh sample for ε

4.6 ROLLER COMPACTI

Compaction is undertaken in t
ing on available equipment, n
product requires compromises
result obtained. Engineering d
in the field, as compared with

The choice of compaction
earthworks costs while achievi
soil. Earthworks design must
available, and any practical co
influence what can be achieve

The soil *in situ* in the borro
particularly if it is a residual
layers of deposition will usua
samples for testing can be a
quality in compacted fills may

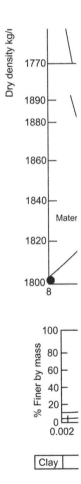

Figure 4.9 (a&b) Comparison of labo
grading curves for two soil

The optimum moisture con
compaction equipment by a pro
field trials as an initial compon
used to optimize equipment sele
moisture condition (see Figure

These include:

- dry clods of soil in a wet r
- shearing due to distortion (
  resulting in shear surfaces
- de-bonding between com
  horizontal direction. This
  addition of loose fill layer:
  permeability,
- poor trafficability of con:
  avoided by proper attentic

The final product must l
fore important that adequate :
construction process.

The compaction character
applying the compactive energ
little resemblance to the comp
is illustrated by Figures 4.9a a
curves for a residual weathered
not prove possible to achieve t
until it was discovered that tl
was 3% wet of that for rolle
permeability and compaction
the fill compacted at roller opti
excavating a shallow (150 mm
engine oil to inhibit evaporatic
the permeability was acceptab
site to "100% USBR laborator
+2%". Figure 4.9c shows the
illustrate the variability of mat
by blending the upper, more w
the less weathered underlying

In the case of soil B, ro
maximum dry density, althou
were almost the same. Howe

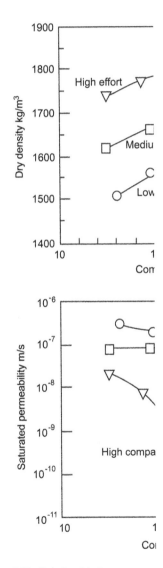

Figure 4.10 Relationship between cor
to water flow for low, me

Figure 4.11 Effect of desiccati

permeability. As the dry unit v
meability decreases sharply an
of optimum. Thereafter, as tł
increases again, slightly, as ind

It is important to realize thɛ
the water content of the soil chɛ
soil may have a particularly de
cracking. As illustrated in Figu
the permeability by an order
eliminated, if the cracks are fc
large. This is also illustrated b

4.8   DESIGNING A COMI

The method for designing a claʏ
permeability, is based on a sɛ
procedure is as follows (USEPɅ

Suppose that the specifiec
separate plot of dry density vɛ
correspond to a permeability ⱱ
in Figure 4.12, in which the cr
content and dry density within

1500 —

1400 —
10

Figure 4.12 Zone for which permeabi
data shown in Figure 4.1(

Strength requirements may also
as indicated in Figure 4.12.

If it is not possible to reach
ability, it is usually possible to ı
a proportion of a clay mineral s
mineral, that occurs as two m
where either sodium or calciuı
bentonite is more expansive aı
it should be noted that ground
soil may be calcium-alkaline wh
more permeable calcium bentor
the chemistry of the seepage wa
of bentonite is not used for the

The bentonite is best mixed
cal mixer before placing, spread
is to be used, a "pulvi-mixer" s
(5 per cent of sodium bentonite
the compacted soil by a factor ɑ

It is particularly important
that is designed to be of low per
sun after compaction. The usual
to be compacted over a compact
When the soil will be exposed
when used as a capping layer, it
desiccation. For example, the ꜱ
single-sized gravel.

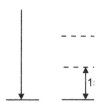

Figure 4.13 Liquid flow betweer

4.9 SEEPAGE THROUGH

The permeability of extensive f
by defects that inevitably occu
shrinkage cracks, poorly mixe(
between layers. Figure 4.13,
stress cracks, interlinked by zo
clay liner. The mechanism of l
the liner and, after some time,
the dye (USEPA, 1989).

It would appear that a mu
than a thin single layer becaus
through faults in contiguous l
limited extent (A in Figure 4.1

Figure 4.14 shows measu
meability. Theoretically, becau
gradient i ($k = v/i$), and i woul(
ity, seepage flow should be in(
more interruption of defects su
thin layer, permeability decreas
ure 4.14 that what was judged
lesser decrease of k with increas
considered to be "excellent" ar
ing the lower bounding line, a (

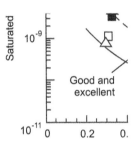

Figure 4.14 In situ measurements

thick liner achieved k = 4×10^{-1}
ness. Thus it has to be conclud
clay liners will considerably out
additional compacted clay (USE

4.10 CONTROL OF COM

In order to set appropriate cont
understood, and a number of ac

- construction requirements i
  pit and compaction water c
  wind, low or high temperat
- moisture conditioning requ
- availability and standards c
- if available, performance d
  specification or optimum cc
- availability and standards c
  tions have been based.

Table 4.1 is a summary of te
can be used as a guide to selec
(e.g., number of m^2 of a chosen
a number of additional factors,

Nuclear meter U

3. In situ strength
In situ CBR C
Penetrometer Ve
Shear vane Fa

4. Permeability in situ
Infiltration from surface Si
as
Drill/auger hole (lateral Si
seepage)
Covered double ring Si
infiltrometer

5. Laboratory tests (on
undisturbed samples)
Constant or falling head C
permeability re
UU triaxial strength Fa
CU triaxial strength Sl

Table 4.2 "As compacted

I	*In situ* dry d
2	*In situ* water
3	*In situ* dry d
4	*In situ* stren;
5	*In situ* perm
6	Laboratory
7	Recipe spec

Table 4.2 lists five parameters

4.10.1 *In situ* dry densit

In principle this is simple and d
time consuming and subject to
moisture/density meters requi

In situ density has traditio[...]
due to its adoption and wide us[...]

4.10.2 *In situ* water cont[...]

Water content can be measured[...]
variability can be assessed relat[...]
that the strength and permeabi[...]
relation to water content, wate[...]
ter. However, the desired proper[...]
density as well as water conten[...]
plemented by measurement of c[...]
trial, the method of compaction[...]
density. Water content could the[...]
would apply only in unusual ca[...]

4.10.3 *In situ* dry density

In section 4.8, a method was de[...]
soil. This requires achieving a r[...]
and permeability, as illustrated [...]
Figure 4.15 shows the statis[...]
dry density that relate to a large[...]
histograms relates to progressiv[...]
that control adequately met the[...]

4.10.4 *In situ* strength

In principle, this is the most effe[...]
ment for performance (see, e.g.,[...]
with maximum dry density. Ho[...]
and b) this may not happen with[...]
control parameters, the acceptal[...]
set. However, it seems unlikely t[...]
75% in Figure 4.16, the strength[...]
say that the air voids $(1 - S)$ in [...]
be no lower than 92%. This v[...]
4.16 to Optimum +3% or abov[...]

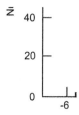

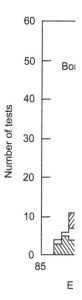

Figure 4.15 Progressive histograms
and improving control a

should be measured with a rap
problems. A variety of rapid si
held or hand-operated vanes or
penetrometers. Difficulties ma
the performance and interpre

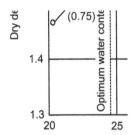

Figure 4.16 Compaction test on clay
(Wesley, 2010b). Figures i

advanced compaction machiner
response of a fill to vibration, i
greatly dependent on water con
the water content changes subse
be used as an essential second c

4.10.5 *In situ* permeabili

This is very effective where perr
for field performance. The grea
is time consuming, very prone
sufficient number of tests to give

Approximate field permeab
ily performed provided that th
construction activities. In this ca
useful guide.

Another disadvantage of j
required. There are no effectiv
that other forms of field tests a
sole control parameter.

4.10.6 Laboratory stren;
in situ measurem

The advantages of laboratory t
can be combined with correlat
compaction control can be achi

imum number of roller passes.
by a minimum coverage "C" v

$$C = \frac{A_f}{A_d} \times N \times 100\%$$

A_f = area of foot, A_d = area of
of C would be 150–200%, tl
the whole area of the layer be
pressure at least once.

(An attempt was made in t
surface of an area of soil by c
that not only is the foot conta
the animals in the second and
predecessors, so that coverage

4.11 SPECIAL CONSIDE
LARGE RATES OF I

In arid and semi-arid condition
of the earth works during a sir
of, for example, diverting a riv
weather, but limits the constru

Placing and compaction oj
ing day. Unless the working ɑ
150 mm thick, may be deposit
This precludes the usual metho
dry density and the compactio
not up to standard. Even if rapi
used, the construction schedule
and recompaction of substand

Because of this, methods (
the correct water content befo
roller passes is used to produc

Large water losses due to
cially if it is hot, and the *in* ɕ

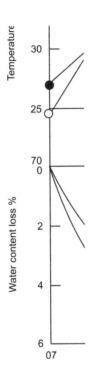

Temperature

30

25

70
0

Water content loss %

2

4

6
07

Figure 4.17 Water content losses fr
conditions. Losses are av

season progresses. As a result, tl
considerably before compactior

The water content is some
area. A few hours may elapse
compaction; in this time a consi
delay results from the practical
rollers and water tankers out o!

Typical water losses from t
the surface of the embankment

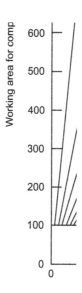

Figure 4.18 Water content contro

The shade and sun temperatur
The water contents represent
loss of water content of 6 perc
the first three hours. Because o
such conditions, to add and m
before compaction. Ideally, the
soil, but this is not practical w
   Using the known capacity
the area of embankment to be
number of full tanker loads tc
a chart is shown in Figure 4.1
on the day and night shifts. Tw
taken from the incoming soil
required compaction water co
the appropriate evaporation-ti

production. Both the supervisin
allow for it in their schedules ar

The particular advantages
conditions and to time constrai

4.12.2 Compactor perfo

Knowledge of compaction perfo
to results which directly reflect
example, field maximum dry d
mined for the combination of s
related to the required number
Figure 4.9).

4.12.3 Testing frequency

The necessary testing frequency
If materials are variable, or if
testing frequency is required tha

Selection of lot sizes (i.e.,
should be based on site conditio
that sampling of a compacted
should take place on a square g
locations that is independent o
grid points from one layer to th
corresponding to 100 to 44 tes
thick compacted layer.

Rapid field tests can be iden
by waiting for laboratory test r

4.13 COMPACTION OF R

Some special characteristics of
paction process is to be unders
optimized. The following char

as well conditions. Many a

by frequent or seasonal rainf
difficult and the characteristic:
the problem of effective comp:

Residual soils are widely
dams and road embankments,
and as impervious compacted
taining smectite or halloysite
volumetric stability, either bec:
ume with varying water conte
smectitic and halloysitic mater
in water-retaining embankmen
are Sasumua dam described b
et al. (1982) as well as number

The parent rock is usually
or a highly faulted sedimentar
Hence the selection of represer
the same reason good control
extremely difficult to achieve.

As examples, Figure 4.19
grading analysis and Atterberg
Note the sharp transition fron
similar variation in lateral exte
a weathered norite clay simila
made to construct a storm wa
variable contact between the r
the light-coloured silty sand ha
variability makes selection of
very difficult to achieve in pra

Drying of a residual soil fr
and physical properties, incluc
soil samples have to be treatec
paction tests are to be at all
laboratory procedure on the
1974) is well illustrated by Fig
soil significantly altered by air
dry density also changed. The

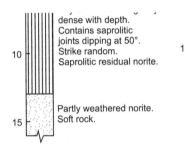

dense with depth.
Contains saprolitic
joints dipping at 50°.
Strike random.
Saprolitic residual norite.

10

1

Partly weathered norite.
Soft rock.

15

Figure 4.19 Variation in vertical dire
weathered norite gabbro

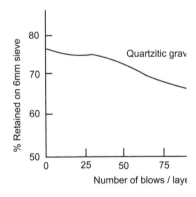

Figure 4.20 Effect of increasing co

which each point represents cor
of the same sample should be
soils, because of the progressiv
during compaction, each point e
of a fresh sample, which is di
sample should never be re-comp
(Gidigasu and Dogbey (1980)) v
size under compaction of a qua
the weathering of granite.

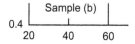

Sample (b)

0.4

20   40   60

Figure 4.21 Compa

Wesley (2010a) produced
weathered volcanic ash believe
content was 165%. As the soi
compactions were carried out
and 30% (that represented air
compaction curves were carri
Wesley does not mention the t
for the water to spread homog
the re-wetted soil. However, as
may have approached that for
had been allowed to cure for s

4.14 THE MECHANICS (
SOILS DURING AN

The mechanics of unsaturated
stand their behaviour:

- during and at the end of con
  by the compaction water c
- in the long term when effe
  climatic conditions, and th
  water content.

During the construction c
at constant water content und
construction the process of co
shows the three dimensional re
ponents of effective stress ($\sigma -$
unsaturated compacted soil (li

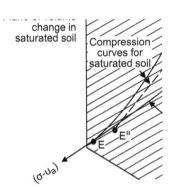

Figure 4.22 Three-dimensional stress-
compacted soils.

As the compressive strain in
disappearing when the soil becc
shows similar data plotted in the
pacted silty clay residual from s
on the compaction water conter
paction water content is increase
compaction water content is inc
compaction water contents.

Figure 4.24 shows that a c
compacted at 0.5% wet of Prc
height of overburden to be place
be reduced to zero. In this instar
hence for the cases illustrated by
tial for swelling if the water cor
However, soils compacted well
lapse when wetted. Figure 4.2.
in the event of either post-comp
(BB″C″), after wetting.

After construction, the wate
equilibrium with its immediate e
structure, the water content wi
soil becomes established. In a se
compacted soil structure, and

Suction (u_a-u_w) kPa

150

200

250

300

350

Figure 4.23 Variation of suction with
silty clay residual from ˅
water content is 16.2%.

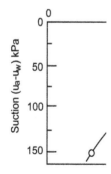

Suction (u_a-u_w) kPa

0

0

50

100

150

Figure 4.24 Variation of suction with
a weathered conglomer
At zero overburden pre

COMPRESSION

When an unsaturated soil, such
undrained conditions, e.g., in a
pore air pressure will rise acco
pressed pore air dissolves in the
of the pore air, being a diffusio
and as it occurs, causes the por
$(\sigma - u_a)$, allows the applied stre
complex reaction between Boyle
the effective stress components

The pore air pressure in a
gauge), but is positive. Figure ∠
sealed base of an AASHTO cor
the mould. The figure shows ho
air into the atmosphere at the t
of the air into the pore water.]
pressure in a freshly compacted s
membrane and immersed under
air, mercury does not dissolve a
were completely impervious. Th
from solution of pore air in the

Figure 4.26 shows changes i
the confining stress on a saturat
simultaneous and instantaneou:
expansion of the soil, the soil co
reduction of the confining stres
by 30 kPa and the soil began to
pore water and u_a and u_w separ
the pore air and water pressures
had approached a steady value.

Calculations based on the :
to the compression of the pore f
predict pore air pressures at the
seems to have been developed b
Hilf (1948).

Figure 4.25 (a) Pore air pressure ca
sample of compacted cl

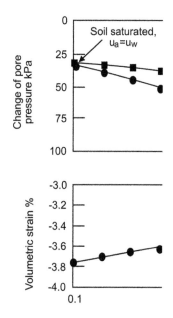

Soil saturated,
$u_a = u_w$

Figure 4.26 Pore pressure changes i
pressure.

The relationship between
expressed as:

$$\Delta u_a = \frac{u_{ao} \cdot \varepsilon_v}{n_a + Hn_w - \varepsilon_v}$$

n_w = volume of water per unit '

H = Henry's coefficient of solu
per atmosphere pressure

$\varepsilon_v(\text{sat})$ is given by:

$$\varepsilon_v(\text{sat}) = \frac{\rho_d e_o (1 - S)}{G_s \rho_w}$$

Equation 4.2 is often (apparentl
be considerably simplified by c
ones. If this is done (Blight, 20
air pressure in terms of the deg
absolute temperature θ in Kelvi
results:

$$u_a = \frac{NR\theta}{(1 - S) \cdot (m_w/S\rho_w) + F}$$

where $N = \{(\text{mass of air})/m_a\}/\{(\iota$
ρ_w = mass density of wate
H = Henry's constant in ʜ

The pore air pressure at which ʜ
$S = 1$, and is:

$$u_a(S = 1) = NH$$

The relationship between pore a
and degree of saturation S is grɛ

Once the pore air pressure
from Figure 4.27, the pore wat
the soil suction. As examples,

$$u_a = +50 \text{ kPa}, (u_a - u_w) =$$

$$\text{(below atmospheric)}$$

$$u_a = -50 \text{ kPa}, (u_a - u_w) =$$

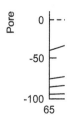

Pore

0

-50

-100

65

Figure 4.27 Calculated variation of
compression of partly s

The following numerical v
are required for calculations o

m_w = molecular mass of water

m_a = molecular mass of air = (

R = universal gas constant = 8

Values of Henry's constant in

Absolute temperatu K
273
278
283
288
293
298

In calculating the curves show
worthy that, at values of S bel
pressures below atmospheric p

The question also arises
opposed to a laboratory test
ure 4.28 shows a series of co
clay core of the Bridle Drift r

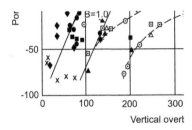

Figure 4.28 Relation between measu...
South Africa.

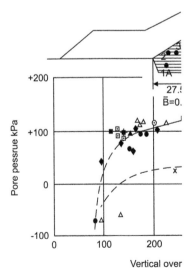

Figure 4.29 Relation between measu...
Zimbabwe.

The soil used for the core was
not as complete as in the case
altitude of 1500 m AMSL, cavi
even more likely. Nevertheless,
negative pore air pressures in a

4.16 SUMMARY

Most of the compaction techno
soils. Differences in the proper
taken into consideration. Bec
variability of the soil will reflec
material may relate both to
attributes. Compaction assess
ensure that specification requi

Field compaction trials are
be controlled as much as possi
is no problem with field trials
it is important to realize that
production. Both the superviso
it in their schedules and budge

The particular advantage:
conditions and to time constra

- Knowledge of compactio
  laboratory test results, to
  equipment. For example,
  imum dry density and o
  compaction equipment, w
  the required number of rol
  results demanded by the d
  dards such as compaction
  necessarily be directly app
  field compaction equipme
- Selection of lot sizes and t

pore pressures. As compac
mechanics of the unsaturat€

- The effects of post-construc
  of the compacted soil and t

Because of the experience c
on applications in semi-arid to
sedimentary and ancient igneous
on residual soils for more comp
volcanic soils in wet climates.

REFERENCES

Blight, G.E. (1962) Controlling e
    Engineering, pp. 54–55.
Blight, G.E. (1970) Construction p₍
    on Large Dams, Int. Comm. on .
Blight, G.E. (1989) Design assessme
    Int. Conf. on Soil Mech. & Foun
Blight, G.E. (2000) Air-water solu₍
    Geotech. Eng. 2000, Bangkok, 1
Bruggeman, J.R., Zanger, C.N. &
    Tech. Mem. No. 592, Washingto
Gidigasu, M.D. (1974) Degree of
    engineering purposes – A review.
Gidigasu, M.D. & Dogbey, J.L.K. (
    residual gravels for pavement co:
    Eng., Accra, Ghana. Vol.1, pp. 3
Hilf, J.W. (1948) Estimating constr
    Soil Mech. & Found. Eng. Vol. 3
Mitchell, J.K., Hooper, D.R. & C₍
    Soil Mech. & Found. Eng. Div. ⁄
Rodda, K.V., Perry, C.W. & Rober
    in a tropical environment. In: Er
    ASCE Geotech. Div. Spec. Conf.
Terzaghi, K. (1958) Design and pe
    Civ. Engrs., Vol. 9, pp. 369–388.

5.1 DARCY'S AND FICK'S

In the middle years of the ninete
that the steady state rate of flo
pressure head in the direction of
in the direction of flow. Darcy's

$$v = -ki, \quad \text{or} \quad k = -v/i$$

where v is the velocity of flow v
k is called the coefficient

The negative sign shows tha
tion of flow. The pressure head
water [N/m³], i.e., [N/m² · m³/N

i, the gradient of the pressu

Thus the units of k are [m/s], i.e
The unit weight of water is
If p is the pressure and z is

$$i = \frac{p/\gamma_w}{z}$$

If equation 5.1 is expressed as:

$$\frac{dm}{dt} = -D_c \frac{dp}{dz}$$

D_c is called the diffusion coeffic
where dm/dt is the mass of flui
unit time, with units o
dp/dz is the gradient of
units of $[\mathrm{N\,m^{-2} \cdot m^{-1}}]$
D_c has units of [s].

Figure 5.1 Observed re

Equation 5.2 is known
compressible or incompressibl
The equation of state for ε

$$pV = \frac{R\theta}{m_a} \cdot m, \quad \text{or} \quad m =$$

where p is the air pressure abc
V is the volume of air,
R is the universal gas cc
θ is the absolute temper
m_a is the molecular mas
m is the mass of air p
K = °C + 273 (e.g., 20°(

It is easy to be misled by ε
proportionality k or D is const
water or air flow. Figure 5.1 is
a saturated soil is not strictly d
a clayey sand residual from we
of water in a triaxial cell. Fig
increase in direct proportion tc
was approximately proportior
Figure 5.2 shows relation
both air and water flow thro
In the case of air flow, the cε
the test on water flow. The re
Figure 5.1. For both fluids, f
pressure gradient, but over a
are reasonably accurate.

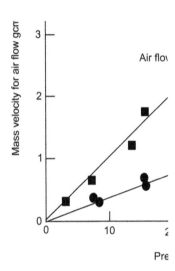

Figure 5.2 Comparison of air and wat
ceramic disc, in terms of m

5.2 DISPLACEMENT OF \

Air can enter a saturated soil onl
to occur, the air pressure at en
capillary forces retaining water
threshold below which the soil
pressure is exceeded, the interco
meability will be established. As
of soil pores will drain and the a
ance is reached between the air
the soil.

In layered soils the thresho
laminations will differ. Air flow
usually parallel to the laminatio
value for flow across the lamin
direction only.

Figure 5.3 Air perm

Figure 5.3 shows the air flo
or air entry pressure for flow a
the laminations. Also, as the a
pores proceeds, the permeabili
In a practical situation this rap
in the soil from reaching the t
the soil may remain permeab
the case in Figure 5.3. The m
steady-state conditions. The so
air pressure before the permeal
soil from which water is being
the form of equation 5.5a (der
the relationship between air p
ratio e, degree of saturation,
5.3 is closely linked to the SW
shown in Figure 5.3 are also sı
is reduced.)

5.3   UNSTEADY FLOW (
SATURATED AND D

Firstly, consider an elemental
through which air at constan
boundaries of the element are i
n and a degree of pore space sa

if m = the mass of air contained

continuity of the air:

$$\frac{\partial m}{\partial t} = M - \left(M + \frac{\partial M}{\partial z} dz \right) =$$

From the equation of state for a

$$m = \frac{m_a}{R\theta}(1 - S)np \, dx \, dy \, dz$$

and hence from equation 5.4a:

$$\frac{\partial}{\partial t}(1 - S)np = \frac{D_c R\theta}{m_a} \frac{\partial}{\partial z}(1 -$$

From equation 5.4b:

$$\frac{\partial}{\partial t}(1 - S)np = k\frac{\partial}{\partial z}(1 - S)n$$

If one is concerned with air flc
is sensibly constant with time a
pores of the soil, n and S can be
b reduce to

$$\frac{\partial p}{\partial t} = \frac{D_c R\theta}{m_a} \frac{\partial^2 p}{\partial z^2}$$

$$\frac{\partial p}{\partial t} = k\frac{\partial}{\partial z}\left(p\frac{\partial p}{\partial z} \right)$$

Equation 5.6a is identical with
consolidation of a saturated soil
is nonlinear.

If equation 5.6a describes,
through dry rigid soils, and th
dition, the large number of estal
dimensional forms) that have be

5.4 UNSTEADY FLOW (

An unsaturated soil that conta
able to air. If the air pressure a
pressures retaining water in th
of water. Any displacement tha
the soil.

Curve A in Figure 5.4 sh
specimen of clay compacted 2%
of saturation of the specimen
amount of pore space availab
Nevertheless, the experimenta
predicted by the Terzaghi theo

Curve B in Figure 5.4 sho
compacted clays saturated by
against a back pressure to mai

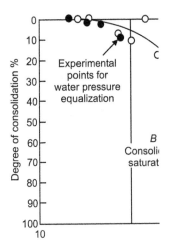

Figure 5.4 Line A: data for unsteady
consolidation of two satu

...nating coarse and fine layers,
sensitivity, small scale measure
ments) tend to be unreliable b
a macro-feature, and if it does, t
will not be correctly simulated
are preferred. However, at the
alternative to measuring perme

There are two basic types of
permeability – rigid walled and
ters consisted of a simple cylinc
the particulate material to be te
was realized that leakage betwee
ments, resulting in apparent per
values. The solution adopted wa
in a triaxial cell where the flexil
the cell pressure, effectively pr
ble walled permeameter. For sp
crushed gravels, or for very lar
punctures in the flexible membr
consisting of a rigid cylinder, li
highly pervious large-particled r
such as a relatively impervious c
can be used to prevent side wall

5.6 OBSERVED DIFFEREI
LARGE SCALE PERM

Day & Daniel (1985) and Dani
measurements of permeability c
and samples were later retrieve
Measurements of seepage rate
of single and double ring infilt
permeameters were made on bl
and also on samples compacted
laboratory were about 100 kPa

did seepage rates from a test pc
resulted in a clay with a perm
same roller without vibration.
gradients of 20 to 100 and effe
the field proved to be between
in the laboratory.

Pregl (1987) has stated th
an index of material quality b
totype lining in the field. The
measured in the laboratory (a
laboratory tests is usually of t
to unity. Also, the Darcy coeff
dient (see Figure 5.1). Howev
field–measured permeabilities

It is apparent from these
permeability measured in the f

- A large area exposed to s
  cracks and more permeabl
- If the Darcy coefficient of
  of different seepage gradi
  field and laboratory value
- A similar remark applies
  effective stress can be exp
  with a low effective stress.

Nevertheless, it is possible
oratory permeability tests, as
ponding tests (shown later in l

All laboratory permeabilit
flexible wall triaxial type per
effective stress was kept at 3 k
stress was the lowest value tha
similar to the effective overbur

Table 5.1 compares the
specimens compacted to the sa

However, it must be noted t
the permeabilities that were con
the literature that discrepancies
appear to increase as the soil be
continuities and defects may pla
volumes of the soil.

5.7 LABORATORY TESTS

The two main types of test for p
gradient test and the falling hea
meability values determined in
behaviour of residual soils. Thi
relatively small size of laborato
geological discontinuities, e.g.,
other relict structures, present in
cracks in a transported profile
law is only partially valid and
considerably with the flow gra
similar flow gradient in the labo

Conventional permeability
compacted soils and more unifo
permeability is determined on b
then possible to estimate the ov
soils. Laboratory tests, unlike
indication of the variation in th
stress. These data are often imp
not available from field tests. Co
apparatus coupled with pore pre
solidation c_v, are particularly us
of permeability at various effect
swelling) and its relation to con

In constant head tests the pe
pore pressure differential (10–2
and by applying Darcy's law wh

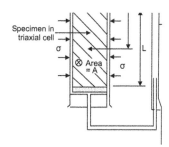

a. Constant head

Figure 5.5 Layout of lab

of permeability can be determ
defining equation (5.1) for Da

$$k = \frac{q_\infty L}{A\Delta H} = \frac{v}{i}$$

where q_∞ = steady state rate o
    A = area of cross-sectio
    $q_\infty/A = v$,
    L = length of sample,
    ΔH = constant differen
    $\Delta H/L = i$.

The layout of the apparatu
which $H - \overline{H} = \Delta H$, and that
For the falling head test, t
the open standpipe of cross-se
rate of flow is $v = a\Delta H/\Delta t$ wh

$$-a\frac{dH}{H} = \frac{Ak}{L} \cdot dt$$

Integrating between times $t = ($

$$k = \frac{aL}{At} \cdot \ln\left(\frac{H_0}{H}\right) = 2.3\frac{a}{A}$$

where $H_0 = H$ at $t = 0$ $H =$

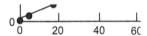

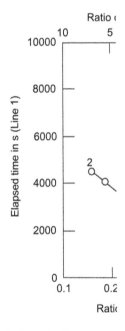

Figure 5.6 Typical results for permea
Note that $\ln(H_0/H) = -\ln($

In less pervious soils or waste r
the steady state value q_∞. It ma
quantity of flow to accumulate.
for less pervious soils Not only i
falling head can be magnified b
area.

Figure 5.6a shows typical re
specimen, in which the cumulati
gradient was 10 kPa over a leng

Figure 5.6b shows results
imens. Here, the ratio of the he
elapsed time t. k can then be c

Table 5.2 shows the range
types of soil. The values are gi
315 360. For example, $1 \times 10^-$

5.8 MEASURING PERME

Permeability to air can also be
However, because air is a con
of air passing through a labor
to be measured at constant pr
falling head principle (Blight,
ply is kept constant and the c
measured.

Differentiating the equatio

$$\frac{\partial m}{\partial t} = \frac{Vm_a}{R\theta} \cdot \frac{\partial p}{\partial t}$$

Hence, from equation 5.2:

$$D_c \frac{\partial p}{\partial z} = \frac{Vm_a}{R\theta} \cdot \frac{\partial p}{\partial t} = D_c \cdot$$

where d is the length of seepa
its superficial area. It is assum
pressure, i.e., reduces from p t
Rearranging:

$$\frac{\partial p}{p} = \frac{D_c AR\theta}{dVm_a} \cdot \partial t$$

Integrating:

$$\frac{Vm_a d}{D_c AR\theta} \ln(p) = t + consta$$

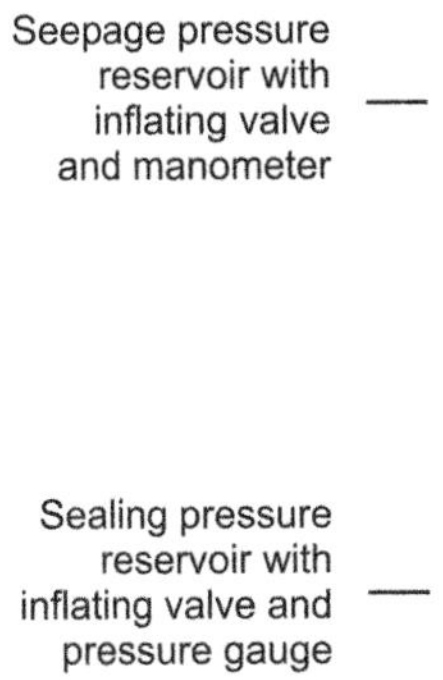

Seepage pressure
reservoir with   —
inflating valve
and manometer

Sealing pressure
reservoir with
inflating valve and   —
pressure gauge

Figure 5

When t = 0, p = p_0 ∴ constan

Finally, the diffusion coeffic

$$D_c = \frac{Vm_a d}{AR\theta t} \ln\left(\frac{p_0}{p}\right) \text{ [s]}$$

The values of m_a and R are give
    With these values insertec
becomes:

$$D_c = 3.48 \times 10^{-3} \cdot \frac{Vd}{A\theta t}$$

Note that

    θ (in K) = °C + 273, e.g.,

If m_a is in [kg m^{-2}], V in [m^3],
    Figure 5.7 shows the constr
out of a length of 150 mm diam

Figure 5.8 Experimental relationshi
coefficient of air diffusior

sides by an inflated latex rub
plate by a flat coil spring. Th
permeameter and the percolat
valve, shown lower right. The
air is measured by means of a

Figure 5.8 shows some ty|
h_0/h or pressures p_0/p (plotted l
is usually linear, but in some c

5.9 METHODS FOR MEA

The most common techniques
form of either constant head or
holes. It is very common to obt
head tests in the drill stem at v
permeability testing can also b
zones for testing, or by installi

Many near surface soils l
opened up with a manual or m
to remain stable during perme
used successfully in unsaturate
methods for drilling in cohesi
boring, percussion-hammer dr
is important that the inside sur

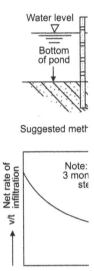

Water level

Bottom
of pond

Suggested meth

Net rate of
infiltration

v/t

Note:
3 mon
ste

Figure 5.9 (a) Layout of ponding test
shielding water level meas
with time after filling pond

loose or remoulded (smeared) m
of the hole or surging it with wa

The most frequently used
divided into two major groups
which extract water. The feed-i
while the extraction tests can or

5.9.1 Permeability from

Ponding tests are suitable for me
and are thus appropriate for obt
the permeability of compacted
for a ponding test are shown in

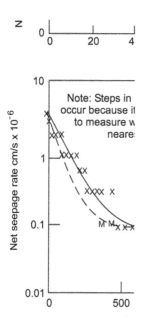

Figure 5.10 (a) Results of ponding
double ring infiltromete

The main pond, on which
edge ponds (or moats) so tha
It is essential that the positio
dimensions of the ponds be re
referred to in Figure 5.10a cc
the four was lined and serve
four water level observation p
the difficulty of measuring sm
in compensating the measured
measuring small changes of wa
of the measurements by causir
wind effects) to change in leng
increase in seepage rate, and v

... ponds ... net infiltration
water level of sealed pond) shou
5.9c and 5.10a and b). Plate
lined evaporation pond. Obvio
by rain precipitation or other n
coefficient of permeability is bas
a steady value.

The single or double-ring in
full-scale ponding test (see Figu
sheet metal rings are set and sea
outer ring serves the same purp
reasonable values of permeabil
than 1200 mm and that of the i

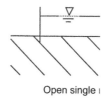

Open single

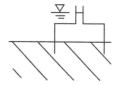

Sealed single ring

Figure 5.11 Open a

Plate 5.1 Set of four moated seepa[
to camera.

The two forms of ring infi
forms. Four variations are ill
because with a low conductivi
of the pond caused by seepage

With sealed rings, howeve
infiltrometers allow flow to sp
pretation of test results. Single
variation: as the system heats
whole system contracts. This
flow is small.

The sealed double-ring infi
forces the infiltration from the
the surface virtually eliminate
of small diameter standpipes r
a standpipe, the flow rate car
inner ring is A, and the cross-
magnified by the ratio A/a. Or
contained in the ring from brea
a particular problem in long-te
have found that the best solut
the ring. This should be at lea:
by pouring plaster of Paris int
the soil surface.

to test with much less scatter in
length of time taken to achieve

5.9.2 Permeability from
(USBR, 1951 & 197

Variable Head Tests

If the water table is close to the s
the rate of rise of water in a bo
suffers from the disadvantage
horizontal flow. Thus the prese
marker or a sandy layer may ha

If the borehole does not in
out using the same principle. N
water and observe the fall of wa
The rate of fall in the water lev
hole when necessary, as initially
suction in the soil around the h
wetted, a true permeability can

The methods of analysis (F
mining the basic time lag T, for
may be employed. The coefficie

k (or k_h if soil is anisotropi

where:

A = cross-sectional area of
F = appropriate shape facto
k = isotropic permeability,
k_h = horizontal permeabilit
T = basic time lag.

Note that Figure 5.12 is drawn
whereas Figure 5.13 is drawn f
by pumping.

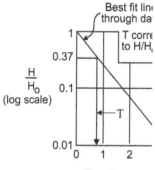

Zon
Defini

Best fit lin
through da

T corre
to H/H,

$\frac{H}{H_0}$
(log scale)

T

Time (linear sc

Best method for cor

Figure 5.12 Calcul

Where the soil is anisotr
obtained from laboratory tests
ing the permeability of the soil
error in a falling head test. (*N
of horizontal to vertical perme

Constant Head Tests

These analyses (Hvorslev, 195
the inflow during a test under
steady state flow conditions a
following steps.

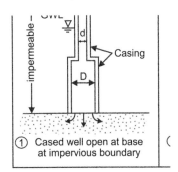

① Cased well open at base
at impervious boundary

i) $A/F = \dfrac{d^2}{D^2 n}\left(\dfrac{\pi n D}{11} + L\right)$

a) $A/F = \dfrac{d^2 m^2}{D^2 n}\left(\dfrac{\pi n D}{11m} + L\right)$

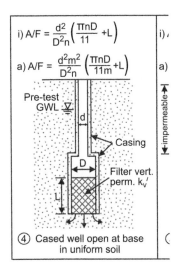

④ Cased well open at base
in uniform soil

Definitions: $k_m = \sqrt{k_v\, k_h}$; $m = \sqrt{k_h / k_v}$

where k_v = vertical permeability
k_h = horizontal permeabi
k_v' = vertical permeability
T is termed the basic tim
See Figure 6.7 for best r

Figure 5.13

Figure 5.14 Typical plot of q

Determination of the stea‹

An approximate value of t
in flow rate with time, as foll
the test hole above the base o{
t denotes the infiltration time,
q versus $\log(1/\sqrt{t})$, or q versu
(Garga, 1988) is shown in Fig
Determination of the effec
For all cases except the p‹
test zone. In the case of packe
zone is adjusted for head loss
additional pressure head suppl
Determine the shape facto

$$k = \frac{q_c}{FH_c}$$

where q_c is the constant flow ı

The analysis of in-situ constan
piezometers is also common p
ibility of the soil and the resulta
is applied. A graph of the rate
(see Figure 5.14). Because 1/✓
approached at $1/\sqrt{t} = 0$. It sho
ous record of the flow for the ‹
flow rate over small time inter
times (from commencement o:
can be plotted against $2/(\sqrt{t_1} -$
of permeability reduces to:

$$k = \frac{q_\infty}{F\Delta h}$$

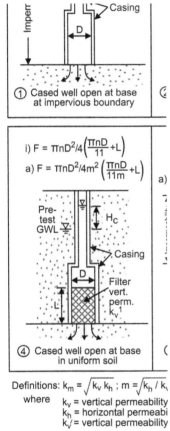

Figure 5.15

where q_∞ = flow rate as t becor
Δh = constant head applied dur
F = shape factor.

The shape factors for cylindric
diameter D, are shown in Figur

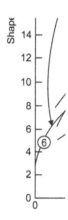

Figure 5.16 Shape f:

5.10 ESTIMATION OF P

Often only a rough prelimin
which case useful results can l
The test-pit is filled with water
several days. The hole must b
evaporation losses and surrou
ning in. The measurements can
magnitude of the permeability.
soak-away test performed on a
test was to see if a low-permea
permeability 0.1 m/y). The resu

The determination of field
widely applied to compacted sc
to work well in soils with perm
This simple method consists o
and length L. The sides of the
necessary to maintain a const
in this case is a three dimensio

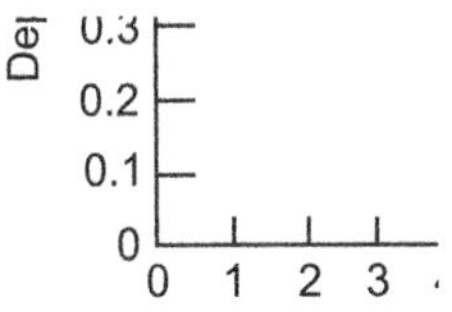

Approximate :
D = (0.75 + 2.

$$A/F = \left(\frac{0.75 + 2.3}{2}\right)^2 \ln \{$$

A/F = 0.45 m   T = 5.5 d

A/FT = 0.082 m/d = 30 r

Figure 5.17 Observed soak-away cu
according to case 5 in Fig

of the flow, the pool is next en
flow rate, q_1 required to maint
ing the two flow rates, the effe
can be eliminated, and the ave
follows:

$$q_{ave} = \frac{q_1 - q}{(L_1 - L)}$$

The range of *in situ* permeabili
pendicularly downwards or ho
expressions:

$$k_v = \frac{q_{ave}}{B - 2H} \quad \text{and} \quad k_h =$$

Figure 5.18 shows the resu
explore possible anisotropy in ç
permeability proved to be cons
base area A of the pit. It could tl
isotropic for all intents and pur

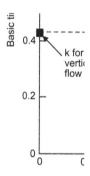

Basic ti

0.4

k for
verti
flow

0.2

0

0   0   C

Figure 5.18 Variation of b

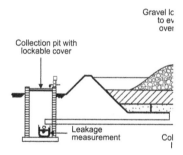

Gravel lc
to ev
ove

Collection pit with
lockable cover

Leakage
measurement   Col

Figure 5.19 Large

5.11 LARGE-SCALE PER

The most realistic way of testir
porate a test pad, or a numbe
can be constructed in accordar
If tests on the pad (or pads) a
prototype and monitored on a
performance of the clay liner.

pad (USEPA, 1989). The dimei
least 25 m × 25 m or 31.6 m × 3

5.12 PERMEABILITY CH/

Despite the enormous influence
for dams, excavations and unde
very limited information on th
macrofabric of a weathering pr
permeability, both laterally and
permeability for various types (
be avoided. Lumb (1975) and
test results from given sites of
of the same order of magnitude
profiles of residual soils presen
(1974), Blight (1988) and others
mineralogy, degree of fissuring,
permeability values from site to
of weathering profiles in igneo
some values of permeability me
in the laboratory.

The methods used to determ
the field and in the laboratory, a
common methods in the field are
boreholes, auger holes and test
review of international practice
preference for in-situ permeabi
test samples to include the micr
field are clearly recognized.

The permeability of a sapr
structure of the material. Most (
termite and other biochannels.
with termite channels in the fot

Table 5.5 Permeabilities of residual :

	Parent rock
Saprolitic soil	Granite
(young residual soil)	Granite
	Granodiorite
	Granodiorite
	Quartz-diorit
	Gneiss
	Gneiss
Mature residual	Granite
	Granite
	Granite

(After Costa Filho & Vargas Jr., 1985).

Table 5.6 Field test method capabilit

	HOLE PREPARATION COST	EQUIPMENT COST	
KEY: — Refers to single test — Refers to stage test RATING: 4 is most favourable 1 is least favourable			
FALLING OR RISING HEAD TEST	3	4	
CONSTANT HEAD TEST	4	3	
PACKER TEST WITH CALIBRATION	4	2	
PACKER TEST WITH PRESSURE TRANSDUCER	4	2	
WELL PUMP TEST, EQUILIBRIUM ANALYSIS	1	1	
WELL PUMP TEST, NON-EQUILIBRIUM ANALYSIS	1	1	

p_...._g ....._ .. ...._.._.._ ...
can be used if the water table is
of field permeability test availab
suitable test for given circumsta
is very appropriate for this purp
to be assigned to available meth
concept and the rating system c
or pond tests.

REFERENCES

Blight, G.E. (1971) Flow of air thr
607–624.
Blight, G.E. (1977) A falling head ;
June, 123–126.
Blight, G.E. (1988) Construction :
Singapore. Vol. 2, pp. 449–467.
Blight, G.E. (1991) Tropical pro
Engineering Geology, British Ge
Brand, E.W. & Phillipson, H.B. (1
International Practice. Hong Kor
Chen, H.W. & Yamamoto, L.D. (19
clay liners. In: Geotechnical and (
Raton, USA, pp. 229–243.
Costa Filho, L.M. & Vargas Jr. E.
Behaviour of Tropical Lateritic ar
Brazil, Brazilian Society of Soil N
Daniel, D.E. (1987) Earthen liners fc
Practice for Waste Disposal, AS(
Daniel, D.E. & Koerner, R.M. (200
ASCE.
Day, S.R. & Daniel, D.E. (1985) H
of Geotech. Eng., 111 (8), 957–9
Deere, D.V. & Patton, F.D. (1971
Mech. & Found. Eng. Puerto Ri
de Mello, L.G.F.S., Franco, J.M.M
soils and behaviour of the foun
Tropical Soils, Singapore. Vol. 1,

Soils: Permeability and Ground

O'Rourke, J.E., Essex, R.J. & Ran
cations to Solution Mining. Wa
Report No. PB-272452.

Pregl, O. (1987) Natural lining ma
Impact of Sanitary Landfills. C

Schmidt, W.E. (1967) Field deterr
and Capillarity of Soils, ASTM

Tan, S.B. (1968) *Consolidation of*
University of London.

United States Environmental Prot
Requirements for hazardous w
89/022, Cincinnati, USA.

US Bureau of Reclamation (1951)
Geology, US Dept. of the Interi

US Bureau of Reclamation (1974)
Denver, CO, US Dept. of the In

Vargas, M. (1974) Engineering pr
2nd Int. Congr., Int. Assoc. of I

Watt, I.B. & Brink, A.B.A. (1985)
Brink, A.B.A. (ed) *Engineering*
Publications. Vol. 4, pp. 199–2

6.1 COMPRESSIBILITY O

Many methods have been used
methods have included the stan
of plate loading tests. Laborat
triaxial compression tests.

All residual soils behave
pressibility is relatively low at
equivalent preconsolidation str
In most cases, the stress rang
pseudo-overconsolidated range

Figure 6.1 shows two typ
andesite lava from two depths
The yield stress is not very clea

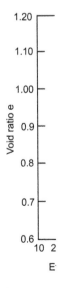

Figure 6.1 Typical oedometer cu

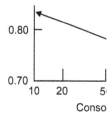

0.80

0.70

10   20   5

Conso

Figure 6.2 Typical consolidation cur
dation pressure and the
mechanical overconsolid;

depth and initial void ratio. Nc
over a relatively small vertical
position of the parent rock. Fig
curve for a residual andesite
preconsolidation stress has bee
lished by mechanical consolid;
transported soil.

The equivalent preconsoli(
particle, or inter-mineral crys
therefore reasonable to expe(
Figure 6.3 shows the variatio:
These data show that because
to increase with depth and the

In a transported soil profi
ratio, and increase with increa
decomposes, the minerals swel
erosion of ultrafine particles,
expansive, it is reasonable to (
be less than the overburden st:
will be less than unity, even th(
yield stress in a consolidation

An estimate of the lateral ;
suction in an undisturbed spe(

20

25

30

Figure 6.3 Variation of equivalent pı
andesite lava.

the overburden stress. If σ'_s is the
equals the suction in the uncont

$$R_s = \frac{\sigma'_s}{\sigma'_{V0}} = K_0 - A_s(K_0 - $$

where σ'_{v0} is the effective overbu
pore pressure in the soil specime
during sampling (σ'_h is the hor
positive,

$$R_s = K_0 \text{ (approximately)}$$

Figure 6.4 shows that there
the profile referred to in Figure
The mean value for the well kno

$$K_0 = 0.9(1 - \sin \varphi') = 0.38$$

while the mean value of K_0 meɛ
there is a strong indication that
confirms the expectation deduce
While data of this sort appear tc
unreasonable to assume that K.
similarly be less than unity, evei

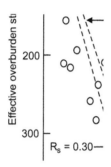

Figure 6.4 Indirect estimates of K

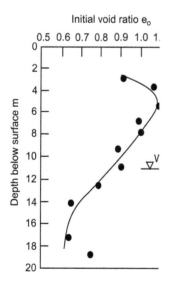

Figure 6.5 Variation of initial void
pressure with depth in pr

Figure 6.5 shows the varia
index C_r in a profile of residual
to decrease with depth (with a
degree of weathering decrease

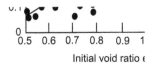

0.1

0

0.5 0.6 0.7 0.8 0.9 1

Initial void ratio ε

Figure 6.6 Correlations between init
 Brummer, 1980). The soli
 Figure 6.5.

three curves suggests that both
Figure 6.6 shows they do. The
shows that the relationships in
the void ratio of the weathered ı
compressibility.

 Similar data are shown in
a profile (similar to the centre
highly expansive clay (or cottoı
friable silty sands which grade ι
9 m). These measurements by]
Figure 6.6.

6.2 THE PROCESS OF CC
    UNSATURATED SOIL

In terms of the unsaturated effec
1965) the processes of heave ι
Figure 6.8 shows the relations
process in an unsaturated soil. I
total stress in the field, a reducı
moisture, will cause the soil to
has a stable fabric or grain strı
soil becomes saturated (at B). T
reduced (u_a does not now exist
unstable grain structure, collap
critical value for the value of (ι
path from D to E and then colla

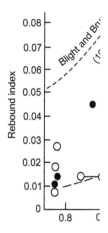

Relations

Figure 6.7 Hall et al.'s (19

may resume along FG, if $(u_a -$
normally.

Figure 6.9 illustrates the
$(u_a - u_w)$ is reduced by increa
adjusted to prevent settlement
plane of zero volumetric stra
experiments have shown that
expansive or collapsing. At con
of the value of $(u_a - u_w)$, as sh

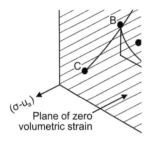

Figure 6.8 Three-dimensional stress-s
constant isotropic load.

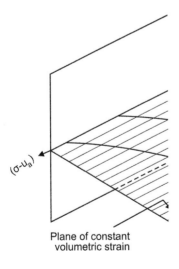

Figure 6.9 Three-dimensional stress-s
partly saturated soils.

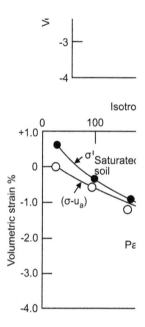

Figure 6.10 Typical curves relating v
(σ − u$_w$) for isotropic cc
also show the compres
21.4%, b: 22.5%.

Figure 6.10 shows detaile
sive soil samples from a pro
compression curves for an un
and the resultant values of (σ −
content. It also shows the con
saturated to an effective stress
strain to changes of σ′, (σ − u
parameter χ as

$$\Delta \varepsilon_v = C\{\Delta(\sigma - u_a) + \chi \Delta ($$

where C is the compressibility.

$$\chi = \frac{65 - 25}{215 - 30} = 0.22$$

and similarly for Figure 6.10b

$$\chi = \frac{51 - 25}{200 - 15} = 0.14$$

6.3 BIOTIC ACTIVITY

This brief introduction would
effects on residual soil compressi
are very common in tropical and
activities may significantly mod
which termite activities may affe
termites either presently or in t
channels, thus materially increa
its compressibility and permeab
been quantified, but it is an effec
and testing soil profiles. It is im
in the soil even though there is r
may have left the area decades
tunnels and tunnelling remain in
 Because termites carry soils
shallow foundations tend to be
foundation, the settlement may
are a few examples of distress c

- Partridge (1989) has repo
 tion of a collapsible soil st
 loaded during a dry period
 water.
- In Johannesburg, one of a p
 a pile cap, started to settle a
 rotated. A 750 mm diameter

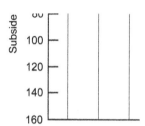

Figure 6.11 Subsidence of various s
Brink, 1985).

cause of the settlement. It
above the food storage cha
structure, about 500 mm i
- In 1974 a precise survey b
  Finnish geodesists (Watt &
  electro-optical and electrc
  1 part in 10 million. The n
  concrete blocks measurin;
  3 m on a yellow to reddi:
  sand, residual from the *in*
  was found that certain of
  had tilted slightly. This sn
  the concrete, hence meas
  distances were set out. A
  The movement continued
  zero and 432 m marks hac
  clearly unable to meet the
  be abandoned. Figure 6.1
  indicating that settlements
  11 years.

There was abundant sur
showed the existence of s
was evidence on surface tl
and depositing it on surfac
the presence of termites oi

...compressibility of residual soils
are also used very extensively. In
control over moisture and stress
in relation to the advantages (th

6.4.1 The conventional p

The plate load test is carried c
the resulting vertical deformatic
in situ, soil disturbance effects
Elastic moduli are calculated fr

$$E = \frac{qB(1 - v^2)}{\rho_0} \cdot I_f$$

In which q is the stress appl
settlement under the plate surfa
and values of I_f are given in Tab
test are usually greater than thos
on specimens hand trimmed fro
moduli derived from plate load t
tests (see section 6.4.3). For a pl
elasticity representative of the s
the plate. In general, the plate lo
of the footing or below this. Plat
crust at the ground surface if the
stratum. Frequently, plate load t
Before conducting a plate load
soaked if it may become wetted
soaking of the soil should be ca
behaviour beneath the water tak
period of time to wet the soil for
This may take several days or e

Test pit

Usually the plate load test is pe
However, elastic half-space the

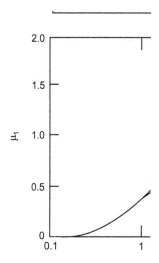

2.0

1.5

μ_1 1.0

0.5

0

0.1   1

E

0

0

Figure 6.12 Elastic settlement calcu
embedment (Christian :

of elasticity, assumes the load
extent. The equation given in]
data, makes these assumption:

If elastic half space theory
should usually be performed. 1
to the plate should be removec
from the edge of the plate, whe
a confined test, the soil surcha
effect of embedment should th
as discussed below.

3.0	l
4.0	l
5.0	2
10.0	2

Notes: 1. Assumptions – pressure q is appl
mass; 2. B = diameter of circular footing ar
pressure; 3. Elastic constants: E = modulus
where λ = numbers given in parentheses
footings, edge of foundation is midpoint of

Plate size and type

Either circular or square rigid
in-place concrete foundations.
reliable size being about 0.3 m.
plate size approaches the actual
can be used to help extrapolate
that soil variability may obscur

Deformation measurement

Plate settlement should be meas
displacement measuring devices
diametrically opposite each oth
Three displacement measuring c
space them 120° apart. The me
to the nearest 0.02 mm. These
reference beam supported on ea
which shows a typical arrangen

Load application

Load is usually applied to the
either a platform supporting a
by helical anchors or tension pi
load cell, proving ring, or other
load accurately. If the jack is sli
false indication of the load. The

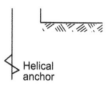

Helical
anchor

Notes: 1. Cond
bottor
2. Test c
the pl
3. Plate
2 dial
refere
the pl
levelir

Fig

to the plate and applied throu;
of eccentric loading.

To obtain good contact bei
is placed between the plate an
alone or mixed with a clean, fii
to minimize bedding errors ar
as thin as practical and be allc
in levelling the plate and apply

Load test

The plate should in general be l
pressure to avoid excessive she
allowing time for the settlemei
next increment. To determine t
as soon as possible after apply
timed to a logarithmic scale, e
the test progresses, plate settle
time. The final test load should
at least 2 hours. After reaching

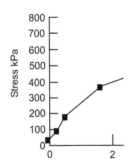

Figure 6.14 Stress-displacement curv
lava soil.

two decrements. Application of
as shown in Figure 6.14. The fir
caused by excavating and settin
based on parameters measured

Primary consolidation settle

For some saturated and/or clay
from a practical standpoint, to ;
practical expedient in testing the
only the instantaneous deform;
instantaneous deformation can
settlement using one-dimension
in the one-dimensional consoli
load test), determine the imme
(i.e., immediate plus primary). I
the total settlement. Plot the ii
settlement, as a function of ap
immediate settlement, expresse

section 6.4.1)

If the relative amount of in
significantly over the range of
each load increment can be co
by the ratio of immediate to t
the consolidation test.

Modulus of elasticity

The modulus of elasticity is cal
performed in the unconfined c
the equation given at the head
elasticity. Use the appropriate
shape of the plate. For a con
Figure 6.12 which considers th

Soil disturbance

In performing a plate load test
the most disturbance because o
This disturbance can be minim
0.5B beneath the plate. Good
modulus of elasticity, however
special instrumentation must b
the plate.

6.4.2 The cross-hole pla

The cross-hole plate load test
plates are used and are jacked
For reasons of greater safely, th
side support in the test hole or
measured, and the lateral comp
of the jack.

The test is very convenien
depths in the same hole. The r
soil, as is the case with the bette

compression does, however, app
at the same depth.

Figure 6.14 shows two str
bearing tests on a weathered an

Based on the assumption tha
modulus of elasticity for the so
(1956).

$$E'_h = \frac{(7 - 8v)(1 + v)\mathrm{Pav} \cdot \pi}{16(1 - v)\rho_h}$$

where Pav is the pressure on th
ρ_h is the movement of th
R is the radius of the pla
v is Poissons's ratio take
E'_h is the elastic modulus

A set of typical results for a set

6.4.3 The screw plate lo

The screw plate load test is a for
depths beneath the surface. The
with a projected area of a full ci
300 mm. The screw plate test is
by hand or machine, using a rel
to the plate. When frictional res
with the same diameter as the p
bottom of the hole. The modul
settlement curve. Valuable info
drained or undrained shear stren
The screw plate test is described
Smith (1987a, 1987b). The prin

Helical screw —
plate t thick

Figure 6.15 Scr‹

Screw plate geometry

The screw plate geometry, illu
following ratios to minimize s‹

where: R = radius of screw p
         b = ¹/₂ of screw pitch
         c = radius of loading
         t = screw plate thick:

A tungsten tipped cutting edge
by abrasion of the soil.

Test reactions

The test plate and loading sha:
by dead load or helical tension
rig, or a heavily loaded vehicle

Screw plate installation

Installation of the screw plate
of the desired test elevation. T
test at a penetration rate per t‹

Use of a rotary drill rig an‹
the screw plate shaft into the l
a test. Using tension anchors t

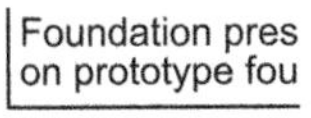

Figure 6.16 Results of screw-plate b‹
were at the same depth ‹

reaction capacity. After augerin
inner loading shaft, which is at
is then advanced using the rotaı

Load test

The load test is conducted by a
plate. A hydraulic jack is used to
proving ring or load cell. The cc
ening of the loading shaft are m
apart. The dial gauges or LVD'
shaft. Relative movement is meɑ
beam which should be at least
from the time the drill rod is rel

The screw plate loading pr
estimated failure load and then
turbance and seats the plate. Tᵥ
applied, and, finally, the screw ı

For low plasticity soils, the l
In residual soils where primary ‹
if possible be left on until at leɑ
has occurred.

Typical stress-settlement cu
Figure 6.16. (Two of the curveɛ
experienced in interpreting *in s*
of a loaded foundation increasɛ
settlements recorded on plates ‹
increase in proportion to their
does not show this trend, as a

where E_{sp} = modulus of elasti
Δq = net increase in av
ρ = measured screw ɪ
R = radius of screw p
λ = a constant depenc
of plate installati

For a deep screw plate located ɡ
a value of $\lambda = 0.65$ can be used
plate. $\lambda = 0.75$ can be used if
of soil. For a shallow screw pl
radii below the surface, $\lambda = 1.($
intermediate depths, values of

6.4.4 The pressuremet

The pressuremeter test offers a
elasticity from the surface dow
does not eliminate, soil disturl
thin-walled tube samples. The
pressure can also be estimatec
pressuremeter test is illustrate
et al., 1984; Mair & Wood 198
pressuremeter pioneered by Cɑ
    Disadvantages of the press
ment which is quite sensitive tɪ
Pressuremeter test results are
test procedures, and the metl
modulus of elasticity is measu
be an important disadvantage
    The pressuremeter test ca
load test. The pressuremeter, ʏ
borehole. The pressuremeter is
the device. Cables and/or pipes
to connect it to instrumentatio
pressuremeter with the soil coɪ

bot

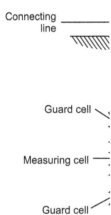

Connecting _____
line

Guard cell

Measuring cell

Guard cell

Figure 6.17 F

is applied to the sides of the h
depending on instrument design

Either gas or water pressu
hence expand the borehole in th
sured indirectly by either determ
chamber or else by deformation
suring the increase of radius of
the change in diameter of the h
approaches, to the horizontal r
pressuremeter.

Hole preparation

The Menard type pressuremeter
residual soils, pushing a thin w
hole has been found to work w

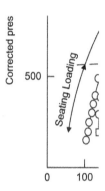

Figure 6.18 Unload – reload cycle in stress release.

should first be expanded, i.e.,
by the foundation at working
reduced to a small seating valu
The reload stress-volume char
used for calculating elastic mc
effects of soil disturbance resu

Equipment calibration

Pressuremeter test results and h
quite sensitive to equipment ca
must take into consideration:

1   **System compliance.** Systen
    ness and also changes in th
    become increasingly impo
2   **Pressure effects.** Pressure l
    resistance to expansion. A
    and the pressure-measurii
    sured at the surface. Figui
    expansion pressure and

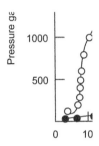

Pressure ga

1000

500

0   10

Figure 6.19 Typical pressur

The membrane has a finite
from the measured expansic
curve and its interpretation

- In phase I, the soil is be
  pressure.
- In phase II the soil com
- In phase III, the soil b
  the limit of volume cha
  membrane.

The pseudo-elastic phase,
calculate the pressuremeter elas

$$E_M = 2(1 + v)V\Delta p/\Delta V$$

where v is Poisson's ratio, usua
$\Delta p/\Delta V$ is the slope of tl
V is the cavity or expan
being measured.

The recycled modulus is typica
loading.

Figure 6.21 shows a set of
depths in a profile of weathere

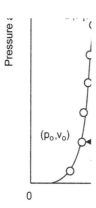

Figure 6.20 Pl

strength, related to the net li
of the soil increase as the dej
correspondingly decreases. As
may vary widely from one hol
pressuremeter moduli measure
profile of saprolitic residual d
stiffness of the soil with dept
the variation from borehole to
borehole.

6.4.5 Slow cycled triax

Multi-stage, slow cycled cons
turbed specimens offer a prac
of residual soils when *in situ*
cycled triaxial tests is that con
high quality thin wall tube san
ples should give the best resul
use of a modulus of elasticity
mated settlements which usuall

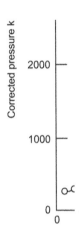

Figure 6.21 Pressuremeter

found that cycled triaxial tests
one-dimensional consolidation

Details of test

The cycled triaxial test usually
three effective confining pressu
reload cycle on each of several s
or a number at the same confin
The effective confining stress fc
value which is used to remove
then determined for the initial p
result of such a test performed
initial bedding stress for the firs

Caution should be exercisec
so as not to damage the usually d
pressures should be selected wl
completed structure. For many

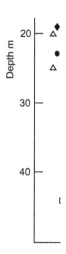

Figure 6.22 Variation of pressureme
test holes.

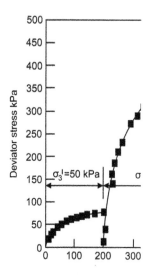

Figure 6.23 "Bedding cycle" triaxial
unloaded and then relo:

---

*N = SPT "N" value.

pressures of 35, 70 and 140 or
triaxial test.

The deviator stress for the ɩ
shear stress $\frac{1}{2}(\sigma_1 - \sigma_3)$ in the sɩ
failure shear stress. An acceptab
past experience for the material
test or tests to assist in deciding

After selecting appropriate ɑ
be plotted using the same scale
nate) axes. For each effective cᴏ
failure circle should be drawn
more than about 30% of the est
ranges suitable for residual silty

At least two complete load-
of the three effective confining p
in a saturated condition if, duriɩ
degree of saturation. A consolic
in effective stress terms.

After initially applying each
as a function of time to determi
Upon reaching the end of primaɩ
may be applied, using an axial sɩ
of pore pressure in the test spec

The problem of deciding on
by means of consolidation theᴏ
1962), but the theoretical solutiᴄ
which they are not (e.g., Blight
was developed. This appears iɴ
nique is needed in Chapter 6, a
measurement. The input data ᴄ
conditions.

Figure 7.24 shows (upper di
B and D for drained triaxial sɭ

20%. (c_v = coefficient of cons(

Figure 7.25 shows the inf
simple chart for which the deg

As far as pore pressure eq
drainage and undrained tests v
The time to failure for 95% de

$$t_f = \frac{1.6H^2}{c_v}$$

Drained and undrained tests
(all-round drains) behave alike

$$t_f = \frac{0.07H^2}{c_v}$$

To convert a time to failu
do a trial test to establish the
particular material being teste(

Modulus of elasticity

For each confining pressure, tl
slope of each unload cycle. Ust
unload cycle is greater than fo
of increase usually becomes le
elasticity is calculated as the
the smaller of the change in l
between displacement measure
i.e., $E = \Delta(\sigma_1 - \sigma_3)/\Delta E$ (axial).

Usually the modulus of el
design. To aid in selecting des
sured elastic moduli for a pai
The modulus of elasticity is p
the horizontal axis. A curve is
obtained for each specimen (i.
fining pressures). Each curve

ison, illustrated in Figure 6.24a

test and oedometer test give co

are used for the plate loading an

should give the lowest disturban

note from Figure 6.24b that the

the same profile has a very simil

SPT results in residual soils may

the correlation is:

$$E = 1.6N \quad MPa$$

A popularly used correlation is

$$E = N \quad MPa$$

which is rather more conservativ

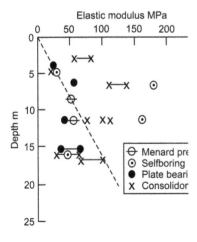

Figure 6.24 (a) Comparison of E valu
profile. (b) Variation of sta
Jones & Rust, 1989).

ual soils provided the Schme
Skempton & Bjerrum (1957) (

Barksdale *et al.* (1982) us
large water tower founded on
based on *in situ* tests could c
based on laboratory tests coul

Working in the same geogr
ment studies of spread footing
oedometer tests, they conclud
normally be about 30% in exc

Barksdale *et al.* have obse
as much as 75% of total settle

The prediction of settleme
above, settlements of structure
the following two techniques:

- one-dimensional consolid;
- elastic theory which uses a
  (section 6.4), or determine

Accurate settlement predi
residual soils. Soil disturbance
and field and lack of knowled
which help to account for actu;
by 30 to 50 percent.

Considerable progress has
behaviour of residual soils. Th
measured *in situ* has given imp
tion within the middle portior
level and thus eliminating end-
results (i.e., higher elastic modu
all specimen deflection. Never
have not proved entirely succe
ence, developed by comparing
verify settlement calculation re

volume changes are climate con
should not be used with highly
50% delayed settlement, unless

6.5.2 Strain influence di

The strain influence diagrams u
sured in model studies and calc
has been observed, in a numbe
with depth more rapidly than ii
for a homogeneous soil. Both i
element analyses show that the
a rigid foundation as implied by
prediction methods. Instead, the
1 times the diameter or width
fore, the strain influence diagra
behaviour than do the common

  The strain influence diagrai
tlement of a foundation is equa
influenced by foundation stress
calculated. Numerical integrati
ing the depth influenced by th

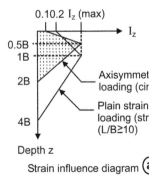

Strain influence diagram (

Figure 6.25 Strain in1

$$\rho = C_1 \cdot C_2 \Delta p \sum_{i=I}^{i=n} \left(\frac{I_z}{E}\right)_i \Delta$$

where $\rho =$ total foundati
$n =$ total number
$C_1, C_2 =$ constants acc
$\Delta p =$ net increase ir
$\Delta z_i =$ variable incre
influence diag
$I_z =$ average strain
$E =$ average modu
increment of (

The correction factor C_1, usec
account for the beneficial effec

$$C_1 = 1 - 0.5\left(\frac{p'_o}{\Delta p}\right)$$

where $p'_o =$ initial effective ov
$\Delta p =$ net increase in eff

The correction factor C_1 shoul
foundations embedded less tha
    The C_2 factor in equation
Secondary compression occur
pressures caused by the applie
from the expression:

$$C_2 = 1 + 0.2 \log_{10}(10t)$$

where t is the time in years aft
    The rate at which seconda
of the soil. As a result, density
ing mica content, and other v
Although equation 6.12 does r
estimates of secondary compr
for residual soils.

about 3B beneath the bottom of
level. Figure 6.26 shows a cros
the soil underlying it. All dimer

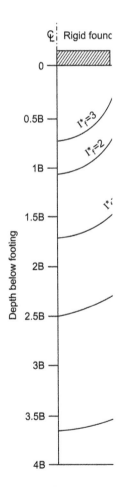

Figure 6.26 Contours of equal vertic
homogeneous soil mass.

$$\rho = 0.10 \cdot C_1 C_2 \Delta p \sum_{i=1} \left(\frac{I_i}{E} \right.$$

All terms in the above equatio
As shown on the contour
caused by the applied loading
from the footing. For many p
than 2B away, can therefore b

Footings at great depth

Table 6.1 can also be used foı
the surface of a deep layer of
used to calculate settlement is

Rectangular foundations: G

Consider the determination of
shape resting on a deep homoş
strain influence diagram meth
ratio, L/B, of the foundation bı
the strain influence diagram in

- the dimensionless maximu
- the dimensionless depth b
  influence factor I_z^0 (max),
- the value of the strain inflı

For a given value of the lε
from Figure 6.27 the value of tl
tion. this figure also shows the
are a function of L/B. The ma
the peak value (z'/B) are obtair
obtained from this figure by tl
rectangular foundation) to obta
lations are carried out using eq

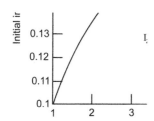

Figure 6.27 Strain influence fact

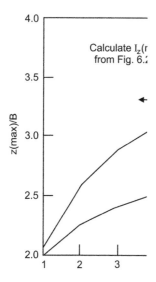

Figure 6.28 Dimensionless maximum
function of L/B, for rectan

Flexible circular, square and
homogeneous deep strata

A flexible foundation is one tha
layer of fill or a thin, lightly rein

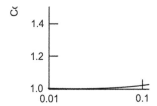

Figure 6.29 Correction factor C_3 fc
modulus of elasticity inc

flexible loadings. A centrally lc
profile with the greatest settler
the edges.

The solutions given in Tab
foundations resting on a deep l
method can also be used for f
rigid foundation having the cc
Then correct the results by mt
the correction factors given in
ment calculation, superpositio
foundation for which influenc

Circular rigid foundation –

The stiffness E of residual soils
is at least partly due to the de
greater with increasing depth.
developed beneath a footing c
homogeneous soil.

To approximate this cond
B underlain by a deep soil strat
elasticity solutions for these cc
diagram approach for a homog
calculated settlement is correc
a circular footing resting on a
soil stiffness at a depth beneatl
this settlement by the appropr

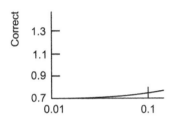

Figure 6.30 Correction factor C₄ for
of elasticity increases line

Figure 6.30, which is simil
when a rigid layer is at a depth (
linearly with depth below the ci
depth of B beneath the foundatic
by the appropriate correction fa

6.5.3 Menard method fo
of shallow foundat

The prediction of settlement is ₁

$$\rho = \frac{(q - \sigma'_{vo})}{9E_M}[2B_0(\lambda_d B/B_0)^c$$

In this equation the first term in
while the second term accounts

where q   = gross average be
        σ'_{vo} = effective overbu
        B_0 = a reference widt
        B   = width or diamet
        α = a rheological fa
           (Table 6.5),
        p_L^* = net limiting pres
        λ_d, λ_c = shape factors de
           (Table 6.6).

Soil Type
Over-consolidated
Normally consolidated
Weathered and/or remoulded

Table 6.6 The shape factor:

	I	
L/B$_o$	Circle	S
λ_d	I	
λ_c	I	

Equation 6.13 applies in cases
footing width B, i.e., $D \geq B$. W
are applied.

The values of α for silt, sa
and the shape factors are tabu

6.6 SETTLEMENT PRED

According to the Brand & Phil
are widely used in residual soi
have been used in Brazil, but bc
widely used in tropical soils. F
bored contiguous piles freque
Bored piles are also used in In
concrete piles are used in Sing
residual soils is the bored cast
piles are also used. The situati

Pavlakis (1983, 2005) ha:
piles from the results of press

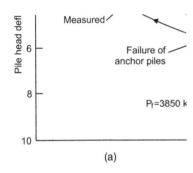

(a)

Figure 6.31 (a) Measured and predict
(b) Measured and predict
one pile cap.

the pile failure loads, he used s
tion 6.5.3), while to predict th
of Sellgren, (1981). Sellgren's n
tion 6.6.1. Figures 6.31a and b
between measured and predict
gle driven cast-in-situ pile, and
pile cap.

6.6.1 Sellgren's method

Sellgren (1981) suggested that
should be taken to be in the for

$$S = \frac{aP}{1 - bP}$$

where S = pile head settlement,
P = vertical load on pile,
α = slope of initial part c
a = tan α,
b = 1/P_f, where P_f is the

Figure 6.32 Load-displacement

Equation 6.14 thus becomes

$$S = \frac{aP}{1 - P/P_f}$$

From the results of many pile t
by equation 6.15

$$a = \frac{4\{1 + \beta/[\theta E_p D_p \cdot \tanh}{\pi D_p \{\beta + \theta E_p D_p \tanh}$$

where D_p = pile diameter,
L_p = pile length,
β = $6E_M/(1 + \nu)$,
θ = $\sqrt{4B/E_p D_p}$,
B = $4.17\ E_M$ for $\nu = C$
E_M = pressuremeter mc
E_p = Young's modulus

For piles with square cross se
where B is the width of the pil

6.7 MOVEMENT OF SHⱯ
RESIDUAL SOILS

The main movement problem
soils arise from seasonal or tiɪ

. . . . , ... , .
of repairs to dwelling houses bui
100 million US dollars per annu
population, but in a country st
population, repair costs represe
or 100 000 fewer people withou
annual increase of population ex
problem.

Solutions to these problems
is referred to the voluminous li
of international conferences on
1965 and 1991. These have sii
wider scope of unsaturated soils
and collapse, used in various c
by Richards (Australia), Gidig
Blight (South Africa), collected
of possible counter-measures wi

6.7.1 Heave of expansiv

Heave is commonly experience
smectitic. Damage to structure:
parts of the world, and is cha
unshaded areas in Figure 1.3a).
sometimes shrinkage) are house
policies usually specifically do
hence the house-owner (either
repairs directly out of his/her p
as to necessitate the demolition
high as 60% to 70% of the pres
guarantee that the damage will
the garden layout, growth of tre
more or less irrigated garden.

As illustrated in Figure 6.8,
remains approximately constan
gains in moisture content. The
surface land-use has changed. F

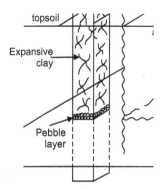

topsoil

Expansive
clay

Pebble
layer

Figure 6.33 Moisture accumulation
moisture supply.

if the land is developed and its
gardens or lawns. All of these
in the soil, thus reducing the
seasonal component to heave.
over part of the year. Hence s
heave will continue during the
of moisture accumulation und
on a slab-on-grade or a stiffen

In an idealized laboratory
pair of tests results for nominal
in Figure 6.34. One specimen
follow the path ABC. If the
total stress, it will swell alon;
A′BC (Jennings & Knight, 1
on a residual weathered shale.
swell will take place even if tl
to 500 kPa), (see BB′ and CC
unloaded in an unsaturated st;
in both Figures 6.34 and 6.35
terms of total stress σ (or σ − ι
A′G, B′G etc. however, are pl
soil will have reduced the suct

0.60

25

Figure 6.34

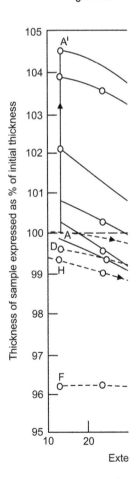

Figure 6.35 Example of d(

15 |‚ ‚ ‚ ‚ ‚ L

0   100

Figure 6.36 Heave of plates buried
Williams, 1991).

Other things being equal,
on the depth of expansive ma
part of the profile is set by the
potentially expansive to an in
non-expansive surface layer.

Deeply weathered mudroc
semi-arid regions. Water table
50 m being quite common. H
tially expansive soil is 30 to .
large (hundreds of mm). Will
approaching 500 mm. At the s
of 10 m below surface (see Figt
material approaches 50 m. Us
seldom exceed 150 mm. Figure
of houses supported on slabs
where the water table was at a
110 mm to 180 mm which exti
not unusually large heave mov

The depth to which seaso
semi-arid zones. There are usu
season of 4 to 5 months follov
dry out to depths of 15 to 20 m
by suction originating from ti
case where the water table un
surface. Beneath an adjacent ei
drawn down, by evapotranspi
the cropland, the depth of pote
plantation were to be felled anc

20

Figure 6.37 Movements of depth poi
profile.

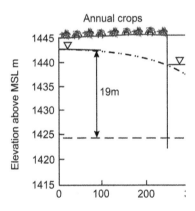

Annual crops

Figure 6.38 Depression of wate

of a wet season often occurs in tł
deeply penetrate the profile dov
(1991), for example, has founc
a depth of 7.5 m, shortly after :
It appears that the heave of th
70 mm, showing that the soil mι
head had heaved extensively anc
pile was progressively exhumed
surface. Figure 6.39 shows the
down the length of the pile. The
appears that there was virtually
depth of 4 m. The portion betwι

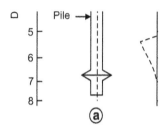

Figure 6.39 Measurements made on
  (b) Load on pile and jac

the length from 5.5 m to 6.7 m
had been sleeved to reduce the
been out of contact with the si
observation has been made by
long dry season. In another cas
Blight (1965) observed an alı
depth point anchored at 5.5 m
this depth.

Engineering solutions to d
For heavy engineering structu:
may be appropriate. For light s
reported considerable success
site and reduce differential mo
another candidate solution. St
for light structures (e.g., Pidge

6.7.2 Prediction of hea

In a comprehensive survey, Sc
ing heave in expansive clay pr
The procedures, in vogue in v
cal methods, usually based on
available to hold increased m
a few applications of effective
effective stress principles, void

$w_0 = $ *in situ* (unswelled)
$PI = $ plasticity index of '

In a later publication, Brackley
equation:

$$\text{Swell } \% = \left(\frac{PI}{10} - 1 \right) \log_{10}$$

where $s = $ suction at centre of ϵ
$q = $ overburden plus fou:

In Brackley's work, the suction
by means of psychrometers or p
to any discrepancy between the
close to the matrix suction.

There must, however, be a
relationships used in geotechni
in conditions and climates that
should not be used unless their

The generalized basis of a r

- The initial and final effectiv
  the soil profile are estimate
- Hence, from measured sw
  calculated, using methods s
- The rate of heave depends
  profile under the changed s
  this appears to be by applyi
  heave is limited by the avai
  void volume governs the ra
  Initial effective stresses can
  measurements using psychi
  both methods), in the labor
  soil can be measured in the
  measurements can be made
  e.g., section 8.5.)

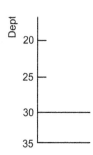

Figure 6.40 Effective stresses in a
channels) estimated fro

A significant difference be
will indicate a significant comp
to take account of the extra h
suction.

Figure 6.40 shows initial i
swelling pressures in a profile re
(Blight, 1984). It will be noted
showed no swelling pressure l
measured values of the swell i
Clearly, from Figures 6.40 and
depth, as does the potential fo

Estimating the final effec
estimating initial effective stres
climate. If the surface will be
contributed to the soil via leak
first suggested by Russam & C
(1965) can be used. The Russa
relates the Thornthwaite mois
450 mm and 3 m below the ce
final effective stresses, it is ne
depth, possibly from the positic
tion is expressed as $-u_w = \gamma_w h$,
then $pF = \log_{10} h$. The Thornt

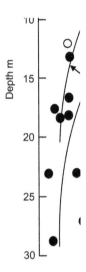

Figure 6.41 Relationship between sv
alluvium-filled channels).

where D = annual infiltration in
       d = annual evapotranspi
       E_p = annual potential ev

In the case from which Fig
station, observations after the
were so many leaks and spilla
profile corresponded to a state
a unit seepage gradient, i.e., th
table.

As stated above, the rate of
in the soil profile under the new
balance for the site. The water l

Rainfall + leakage − runc

= infiltration into soil

pF 2.5 ‒ ‒ ‒ ‒ ‒ ‒ ‒ ‒ ‒ ‒

◄─── l₁

-60   -40

Figure 6.42 Russam-Coleman/Aitch
surfaces to Thornthwait

In this equation, leakage w
and leakage from plumbing an
leaks from cooling tower pond
to estimate an accurate water t
be obtained, if necessary, by n
agreement on numerical desigr
the subject).

Once the rate of infiltratio
occur can be calculated from tl
air-filled pore space in the pro
Where the lateral extent of a
depth of expansive material, v
surface and then by lateral flc
of the time versus heave curve
typical ogival time-heave curv
of expansive soil was initially ;
also shows the rise of ground
the area subject to heave has l;
depth of expansive soil and th
which water can accumulate ii
migrating vertically downwar
this situation, it is assumed th;
wetting front and hence that
expansive soil is usually loca

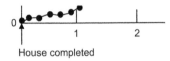

House completed

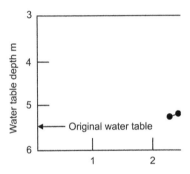

Water table depth m

Original water table

Figure 6.43 Ogival or S-shaped time-
                the soil as a result of a cl
                of seasonal variations car
                table level.

resulting time-heave curve is cc
This is illustrated by Figure 6.4‹

- the simplified characterizati
  outputs and including the v
- the resultant concave-up ti
  tainties, this is shown as a z

Figure 6.45 shows a selectio
to by Figures 6.40 and 6.41 tha
starting at the beginning of year
starting 2 years earlier as the h
Figure 6.45 also shows the larg
point to point over a large site.

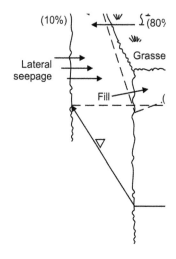

(10%) Σ (80%

Lateral
seepage

Grasse

Fill

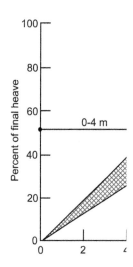

Figure 6.44 (a) Elements of a water
curves for site water ba

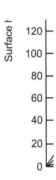

Figure 6.45 Time-surface heave relati⌐
station.

6.8 COLLAPSE OF RESID⌐

The phenomenon of collapse se⌐
are loess or loess-like soils usua⌐
been lightly cemented at the p⌐
consists of unusually highly wea⌐
as granites, that contain a large ⌐
and loss of mineral material, th⌐
high void ratio and an unstable

Ancient wind-blown sands

Extensive areas of Southern Afr⌐
Quaternary age. Because of char⌐
that were deposited under deser⌐
e.g., as Kalahari semi-desert gr⌐
the present moister climate, the⌐
few per cent of silt and clay. Or⌐
their structure has become coll⌐
1961). Figure 6.46 shows Knig⌐
windblown sand. Figure 6.47 (N

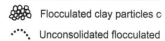

Flocculated clay particles c

Unconsolidated flocculated

Figure 6.46 Knight's (1961)

Co

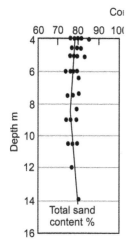

Figure 6.47 Variation of sand, silt a
McKnight, 1999).

percentages of sand, silt and cl
Mozambique. This shows tha
material vary from 92–100% a
varying from up to 8% towarc
leached weathered granites m
that increased annual rainfall c
granite and therefore a higher

The characteristics of a col
sive soil in Figures 6.8 and 6.9,
illustrates the considerable str
fining stress of $(\sigma - u_a) = 15 \, \mathrm{kI}$

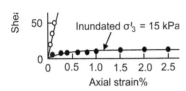

Figure 6.48 Triaxial shear tests on cc
deviator stress vs axial st

the effective stress fell to 15 kPa
that the soil actually swells by a
stress (1.5 kPa). If it is wetted w
relatively sudden collapse settle
strength and be relatively incom
loses strength and becomes col
water content.

Figures 6.49a and b show tl
of an unsaturated collapsing sai
sand after inundation, plotted
sand and σ' for the saturated sai
unsaturated and saturated is ve
similar data for an expansive cl
for the saturated soil do not lie l
the unsaturated soil must obey
cannot be evaluated by applying
and shown in Figures 6.49a ani
the unsaturated collapsing sand
$\Delta(u_a - u_w) = 0$ and equation (6.

$$\Delta\varepsilon_v = C\Delta(\sigma - u_a)$$

Thus the compression of the uns
and χ is either zero or does not

Although collapse cannot t
tions, collapsing soils behave as
after the collapse takes place (se
soils, collapse may take place p

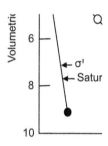

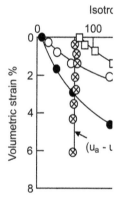

Figure 6.49 (a and b) Typical curve
$(\sigma - u_a)$ and $(\sigma - u_w)$ fc
file. The diagrams also s
contents were: a: 8.3 pe

at all. An example of this is sh
example had been loaded at a
settled without collapse. If, on
tent and was then subsequentl
than 100 kPa, and collapsed if
spectrum between expansive a

The amount of collapse s
of the soil and its water conte
tion loads of 100 to 300 kPa,
seem common, while settleme

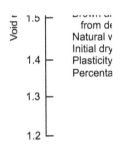

Figure 6.50 Reaction of a loose clayey
content, and after saturat

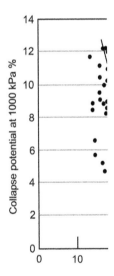

Figure 6.51 Collapse potential vs degı
shown in Figure 6.47.

Figure 6.51 shows data collectec
to the collapse potential (at 1C
ure 6.47. The collapse potentia
of 16% to almost zero at 32%.

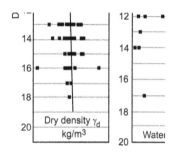

Figure 6.52 Variation in physical par
(McKnight, 1999).

Figure 6.52 (McKnight, 1
dry density, water content, voi
shown in Figures 6.47 and 6.5
on average, almost constant w
If the values $\gamma_d = 1700 \text{ kg/m}^3$
particle solid relative density (
content at saturation would tl
Figure 6.52, the profile was far
start at 6m because site levellir

Predicting collapse settlem

The amount of collapse that w
tests (e.g., Jennings and Knigh
followed by wetting that is lil
relatively little attempt to imp
problem of collapse is wide-sp

Combating the effects of c

Structures most likely to be affe
e.g., roads, housing and slab-o
been used with apparent succe
on the reasons for success of t
assume that collapse will occur

....

observation that traffic loading
to produce a densification to de

The practices used to achiev
the designer or contractor. The
been tried:

- Removal and compaction.
  and it is then re-compacted
  of treatment can usually on
  house or a small building. I
  of these sands, it is often (
  density.
- Densification by rolling or
  been used, e.g., vibrating sı
  nal) rollers, with or withou
  dynamic compaction has a
  been very variable. An inte
  van Alphen (1980) who co
  profile with a heavy vibrati
  penetrated to a depth of 3
  motorscraper having wheel
  block dropped from heights

  - none of the methods pr
    400 mm;
  - initially the roller and th
    breaking down its fabr
    measurable, but small (
  - there was no difference
    and
  - ponding loosened the s

  Weston (1980) in a compre
sands for road construction has
sands *in situ*. Figure 6.53 repro
effect of compacting a section c
roller. Although some densifica

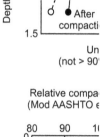

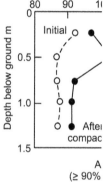

Figure 6.53 Results of rolling trials o
1980).

in the collapse settlement of t
results for a second section of
in reducing the collapse poten

- The maximum relative con
  90% of Modified AASHT
  compactability is strongly
- There is no evidence to sug
  be greater than 90% of M
  other words, the compact
  disappointing were satisfa

Strydom (1999) reported a
clayey sandy silt in which su

0.6 ⌐ ⌐
0   10

Figure 6.54 Settlement vs ni

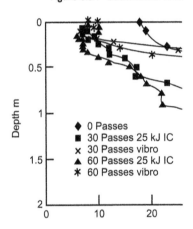

Depth m

- ◆ 0 Passes
- ■ 30 Passes 25 kJ IC
- ✕ 30 Passes vibro
- ▲ 60 Passes 25 kJ IC
- ✱ 60 Passes vibro

Figure 6.55 Variation in DCP

penetration of a cone penetrom(
An array of rollers or compacto

    12 ton vibrating roller,
    4-sided single drum impact
    5-sided dual drum impact c
    5-sided dual drum impact c
    3-sided dual drum impact c

(Energy per blow $=$ Wh [Nm $=$
    Figure 6.54 shows the settl(
the various compactors. There
were vastly superior to the vibr
relatively little to choose betwe(
    Figure 6.55 shows the re
(DCP $=$ Dynamic Cone Penetrc

Aitchison, G.D. & Richards, B.G
ment subgrades throughout Au:
Beneath covered Areas. (G.D. /

Baguelin, F., Jezequel, J.F. &
Engineering. Aedermannsdorf,

Barksdale, R.D., Bachus, R.C. &
Eng. & Const. Tropical & Re
Published, American society of

Bishop, A.W. & Henkel, D.J. (19
London, Edward Arnold.

Blight, G.E. (1963) The effect of n
shear strength of soils. *Laborat*

Blight, G.E. (1965) The time-1
G.D. Aitchison *Moisture Equil*
Butterworths, Sydney, Australia

Blight, G.E. (1967) Effective stress
Div., ASCE, 93 (SM2), 25–148

Blight, G.E. (1974) Indirect deter
Found. Conf. Subsurface Explo
Henniker, USA. pp. 350–365.

Blight, G.E. (1984) Power station
in Geotech. Eng., St. Louis, US

Blight, G.E. & Brummer, R.K. (1£
J. S.A. Instn. Civ. Engrs., 22 (1(

Blight, G.E. & Lyell, K. (1987) Lc
the geotechnical effects of wate
Eng., Dublin, Ireland. Vol. 1, p

Blight, G.E., Schwartz, K., Wet
by flooding-failures and succe:
pp. 131–135.

Brackley, I.J.A. (1975) *The Inter-*
Unsaturated Soil. PhD thesis, U

Brackley, I.J.A. (1980) Prediction
Africa Soil Mech. & Found. En

Brand, E.W. & Phillipson, H.B. (
Scorpion Press.

Bycroft, G.N. (1956) Forced vibr.
on an elastic stratum. *Phil. Tra:*

Christian, J.T. & Carrier, D. (197
Eng. Div., ASCE, Vol. 96, No.

Jennings, J.E. & Knight, K. (197~
   additional settlement due to colla
   Found. Eng., Durban, South Afr
Jones, D.L. & van Alphen, G.H. (1
   Soil Mech. & Found Eng., Accra
Jones, G.A. & Rust, E. (1989) Fou
   Int. Conf. Soil Mech. & Found. i
Knight, K. (1961) *The Collapse (*
   University of the Witwatersrand,
Knight, K. & Dehlen, G.L. (1963)
   Reg. Conf. Africa, Soil Mech. &
Mair, R.J. & Wood, D.M. (1987) Pr
   Testing. CIRIA Ground Eng. Rep
McKnight, C.L. (1999) The strat
   lapsible residual soils on the
   Blight, G. & Fourie, A. (eds) (
   pp. 633–645.
Menard, L. (1965) The interpretatio
Moore, P.J. & Chandler, K.R. (1
   Melbourne. *5th Southeast Asian*
Partridge, T.C. (1989) The significa
   in transported quaternary soils. Ii
   Research. Rotterdam, Balkema. j
Pavlakis, M. (1983) *Prediction of l*
   Tests. PhD Thesis, University of t
Pavlakis, M. (2005) The Menard pr
   In: Gambin, M.P., Mangan, L. &
   de l'ENPC/LCPC. pp. 100–118.
Pellissier, J.P. (1991) Piles in deep r
   Eng., Maseru, Lesotho. Vol. 1, p
Pidgeon, J.T. (1980) The rational de
   Conf. Africa Soil Mech. & Foun(
Russam, K. & Coleman, J.D. (1!
   conditions. *Géotechnique*, 11 (1)
Schmertmann, J.H. (1955) The und
   pp. 1201–1227.
Schmertmann, J.H. (1970) Static
   Mech. & Found. Div., ASCE, 96
Schreiner, H.D. (1987) *State of the i*
   and Road Res. Lab.

6th Pan Amer. Conf. Soil Mech.

Strydom, J.H. (1999) Impact comp
    and methods of integrity testin
    for Developing Africa. Rotterd:

Tromp, B.E. (1985) *Design of S.*
    Johannesburg, South Africa, Sc

Wagener, R.v.M. (1985) Personal

Watt, I.B. & Brink, A.B.A. (1985
    In: Brink, A.B.A. (ed.), *Engin.*
    Building Publications. pp. 199–

Weston, D.J. (1980) Compactioi
    Mech. & Found. Eng. Accra, G

Williams, A.A.B. (1975) The sett
    Reg. Conf. Africa Soil Mech. &

Williams, A.A.B. (1980) Severe he
    Soil Mech. & Found. Eng., Acc

Williams, A.A.B. (1991) The extr
    buildings and roads. *10th Reg.*
    Vol. 1, pp. 91–98.

Willmer, J.L., Futrell, G.E. & La
    ual soil. *Eng. & Constr. in Tre*
    Honolulu, USA. pp. 629–646.

Zeevaert, L. (1980) Deep foundat
    South East Asia Conf. Soil Eng

7.1 BEHAVIOUR AND DII
FROM TRANSPORTE

The selection of appropriate st
and failure are important steps
soils. In countries with widesp
the laboratory and in the field n
and shear box tests in the labor
loading test in the field (Brand &
of the difference between transp
ration and handling of the spec
A knowledge of the genesis of
strength will enable both engine
ciate the peculiarities of these 1
and will thus facilitate the selec

Residual soils develop a pa
bution *in situ*, which may make
The latter develop their fabric a
history after deposition. In the
the stress-strain behaviour and
discussed.

7.1.1 Factors affecting s

Table 7.1 lists special features en
sible for the differences in stress
soils. (Also refer to Chapter 1 a

Stress history

After deposition, transported so
due to increasing depth of buria
or all of the overburden is remov
In the case of clays, which are de
after deposition almost fully det

Bonding	Can be an im strength mos and/or cemer intercept and can be destrc
Relict structure and discontinuities	Develop from features in pa bedding, flow joints, slicken:
Anisotropy	Usually deriv(structure, e.g structures
Void ratio/density	Depends on s process, large history

Residual soils are formed b
of chemical processes (e.g., le̟
is mainly a weakening proces
may also cause some vertical ̟
from voids forming in the alteɪ
in situ stresses which modifie̠
progressively weathering mate
structure of residual soils to b
state of stress. The effect of pas
formation will usually be small
(e.g., via fractures or shearing
Most saprolitic or lateritic
A-parameter during shear is u̟
zero or close to zero at failure.
shear are usually relatively uɪ
ure 7.1 shows the variation of t
undrained triaxial compressio̠
of Figure 1.13. Note that the ̟

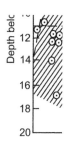

Figure 7.1 Profile

zero and the mean value is less
as the degree of weathering dec

Grain/particle strength

Weathering produces soil partic
variable degrees of weakening.
variability in crushing strength t
cle size distributions are less me
because they depend on the degr
sample preparation.

Bonding

One of the characteristics of a re
These bonds represent a compo
effective stress and void ratio o
ported soils (soft and stiff cla)
geologic age, i.e., when bonds
applications these bonds are u:
strength or compressibility.

Possible causes for the devel

- cementation through the dej
  at particle contacts in the p
- solution and re-precipitatio
- growth of bonds during the
  1.7 for information on pedo

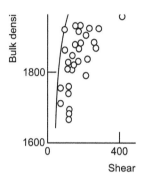

Figure 7.2 Relationship between un
and bulk density for satu

The strength of bonds is v
weathering processes. But it sh
ple can scarcely be handled wit
and stiffness which can have a
the soil *in situ*, especially at sh

Figure 7.2 illustrates how t
progressively weakened as wea
It shows how the shear streng
The bonds in the partly weath
but those in the completely we

Interparticle bonds can be
to consider in sampling, samp
stages of shear testing the type
become progressively non-unif
or fully destroyed, resulting in
structure may be partly destr
confining stresses to a test sp
carefully and incrementally app
less than the undisturbed *in sii*

Remoulding of a residual s
duces a "de-structured" soil in

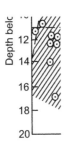

Figure 7.1 Profile

zero and the mean value is less
as the degree of weathering dec

Grain/particle strength

Weathering produces soil partic
variable degrees of weakening.
variability in crushing strength t
cle size distributions are less me
because they depend on the degr
sample preparation.

Bonding

One of the characteristics of a re
These bonds represent a compc
effective stress and void ratio o
ported soils (soft and stiff clay
geologic age, i.e., when bonds
applications these bonds are us
strength or compressibility.

Possible causes for the devel

- cementation through the dep
  at particle contacts in the p
- solution and re-precipitatio
- growth of bonds during the
  1.7 for information on pede

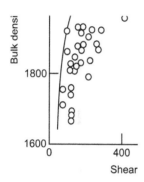

Figure 7.2 Relationship between un
and bulk density for satur

The strength of bonds is v
weathering processes. But it sh
ple can scarcely be handled wit
and stiffness which can have a
the soil *in situ*, especially at sh

Figure 7.2 illustrates how t
progressively weakened as wea
It shows how the shear streng
The bonds in the partly weath
but those in the completely we

Interparticle bonds can be
to consider in sampling, sam
stages of shear testing the type
become progressively non-unif
or fully destroyed, resulting in
structure may be partly destr
confining stresses to a test sp
carefully and incrementally ap
less than the undisturbed *in sit*

Remoulding of a residual s
duces a "de-structured" soil in

residual soil (see Plates C19 anc
and drilling. Test specimens witl
surfaces of weakness, even if th
of maximum shear stress.

Many authors have conclud
almost entirely by their inherit
matrix material contained betv
degree of weathering and also b
(see Figure 1.10a). However, th
predominant orientation and fre
of stress application, and to th
examples of the truth of this stat
Irfan & Woods (1988) and Lo ɛ

In certain cases, it may be pc
mapping the saprolitic discontir
surfaces either *in situ* or in the la
relatively infrequent, occur in a ɪ
a jointed rock mass. In other cas
as a regular soil anisotropy. Fc
idealized in terms of anisotropy
to take a theoretical account c
(1989). In many cases, however,
their effect on soil mass strengtl

Stress anisotropy

As a result of stress anisotropy
depends on the direction of the ɪ
soils, stress anisotropy is directlʏ
history of the deposit. With resic
been inherited from the fabric c
also play a role. This applies paɪ
where mica is present. Plate-lik
can become oriented during th
Such surfaces may develop *in sit*
accompanying soil genesis, but ɑ
and C20). Due to the randomnɛ

which increases with density (s
The void ratio also influences

7.1.2 Effects of partial

Because of climatic conditions
regions are often deep. Evapo
Figure 6.38). This leads to deep
frequently exist in an unsatura
pressure will usually approxim
will be sub-atmospheric, i.e., n
soil. This negative pore water p
of effective stress, or in other v
stress.

As introduced in section 1

where u_a = pore air pressure,
$\quad\quad\quad u_w$ = pore water pressu

The equation for the shear stre
the two stress state variables (σ

$$\tau = c' + [(\sigma - u_a) + \chi(u_a - $$

where χ is the dimensionless B

$$\text{or } \tau = c' + (u_a - u_w) \tan \varphi$$

(Fredlund *et al.*, 1978)
where $\quad \tau$ = shear strength,
$\quad\quad\quad c'$ = effective cohesion
$\quad\quad\quad \sigma$ = total normal stre
$\quad\quad\quad \varphi'$ = effective angle of
$\quad\quad\quad \varphi^b$ = angle by which c

Equation 7.1a can be written a

$$\tau = c' + (\sigma - u_a) \tan \varphi' + ($$

obtained from a set of measurem
an average value for a particular
shown in Figure 7.3 gives speci
constant.

Equations 7.1a and b are a
When a soil becomes saturated,
sure and equations 7.1a and b
Hence there is little difference
no advantage in using the one in

Figure 7.3 shows the relatio:
according to equation 7.1a whil
tion 7.1b. Figure 7.3a shows a ty
soil in which u_a and u_w have l
method used to determine the B
plotted in $\frac{1}{2}(\sigma_1 + \sigma_3) - u_a$, $(u_a -$
of a line such as A'A to the dire
Fredlund and Morgenstern's φ^b
parameter χ in equation 7.1a ar
ship was calculated for an idea
spherical particles (see section
interpretation via equation 7.1b

The terms χ or φ^b must be
Blight, 1963; Ho & Fredlund,
in a range between 0 and 1, bi
Figure 1.14, Blight, 1967b). Tl
from 0 to φ'.

In the case of an unconfined

$$\tau = c' + \chi(u_a - u_w)\tan\varphi', \ ]$$

$$\chi\tan\varphi' = \frac{\tau - c'}{u_a - u_w} = \tan\varphi^t$$

and if $c' = 0$, then at failure:

$$\tan\varphi^b = \frac{\tau}{u_a - u_w} = \frac{\frac{1}{2}(\sigma_1 -}{u_a -}$$

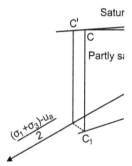

0 ↙
0 ⟶

Satur

C'

C

Partly sa

$\dfrac{(\sigma_1+\sigma_3)-u_a}{2}$

C₁

Figure 7.3 Experimental determinat
tests (Bishop & Blight, 19

Figure 7.4b shows experil
for a series of unconfined comp
as well as the corresponding re
illustrates that the usually poss
    The evaluation of soil wat
particularly important with sl‹

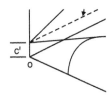

(a)

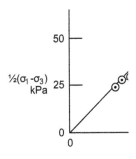

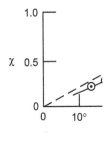

(b)

Figure 7.4 (a) Morgenstern & Fredlu
state variables for an unsa
$\frac{1}{2}(\sigma_1 - \sigma_3)$ and $(u_a - u_w)$
corresponding relationship

the test approaches that
well designed and robust,
However, large scale field
consuming. Because of the
number of tests, and know
2   A large number of small sc
in situ tests could include
and empirical or semi-em]
penetrometer tests. Suitabl
triaxial compression test a
are vane tests using vanes c
penetrometer tests with m
laboratory, shear box test
can be used to explore the

This approach has the advant;
erally and with depth, but wit
values for design or analysis.
    A number of workers, e.g
that for stiff materials containi
over-estimate soil mass strengt
of small scale shear strengths a
    This is simply because the l
strengths of the discontinuities
(Blight, 1969) shows a compar
ods on a lateritic residual we
from a sliding failure through
material. The comparison illu:
ure 7.5 is quite characteristic c
difference between the undist
ticularly marked. The undistu
saprolitic discontinuities while
cially produced fissure surface
of the soil in mass, makes it cle
by the strength along discontir.
the strength measured in small
given in what follows.

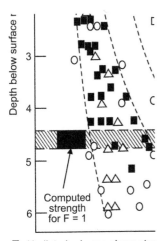

□ Undisturbed vane shear stre

■ Remoulded vane shear stren

Figure 7.5 Comparison of various sr
strength-in-mass calculated

The effect of the spacing of
strength is further illustrated k
measured strength for a specime
and less realistic as its size decre
that is 2 or more times the spacii
realistic. These observations are
which shows the considerable e
residual from a vesicular basalt
and 7.7 that the strength of a s
up to 5 if too small a specimen
that the size effect is more pronc
3.5 for fissured soil) and reduce
350 kPa (the strength ratio for a
1.5 for fissured soil). This show
of voids or partings and that th
similar defects to the "non-intac

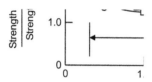

Strength / Strength

1.0

0

0   1.

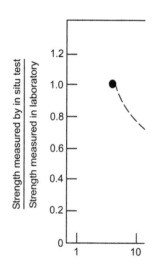

Strength measured by in situ test / Strength measured in laboratory

1.2

1.0

0.8

0.6

0.4

0.2

0

1   10

Figure 7.6 (a) Influence of the ratic
laboratory tests on Lond·
clay (Lo, 1970 and Marsla

Garga's conclusions on th
of a residual soil in mass are w

"1   Discontinuities and fi:
      soils. This behaviour
      clays of sedimentary a
2   The drained strength (
      has been found to be 1.

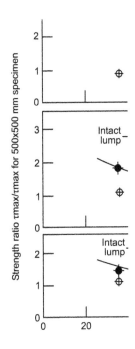

Figure 7.7 **Effect of specimen size**

triaxial samples in the
effect of sample size ha
vesicular residual soil.
3   The frictional compone
variation with sample s
strength with size may
4   The limited data sugges
field strength of fissure
intercept from the stren

*(It should be noted that there a
by [ ]. As the results clearly sho
the author must have meant to

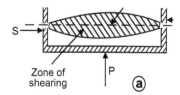

Zone of
shearing

P

(a)

Figure 7.8 Stress system applied in c
through point A and tang

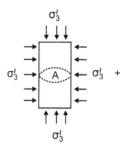

σ'_3

σ'_3 A σ'_3 +

σ'_3

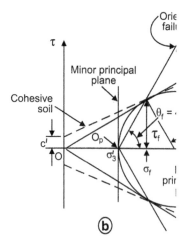

Orie
failt

τ

Minor principal
plane

Cohesive
soil

c'

O

O_p

σ'_3

$\theta_f = $

τ_f

σ_f

prir

(b)

Figure 7.9 Stress system applied in ti
and orientation of princij

Direct shear test

Advantages
- Relatively simple and quick to perfo
- Enables relatively large strains to b
  thus the determination of residual
- Less time is required for specimen
  and drainage because drainage path
  (half specimen thickness)
- Enables shearing along a predeterm
  (e.g., plane of weakness, such as rel

Disadvantages/Limitations
- Drainage conditions during test, esp
  pervious soils, are difficult to contr
  only drained tests are possible. She
  place within a zone (Figure 7.8a)
- Pore pressures cannot be measure
- Stress conditions during the test ar
  indeterminate and a stress path car
  established. The stresses within the
  are non-uniform. Only one point c
  in a diagram of shear stress τ versus
  σ, representing the average shear s
  horizontal failure plane. Mohr's str
  only be drawn by assuming that the
  plane through the shear box is the
  failure plane. During straining the d
  principal stresses rotates
- Shear stress over failure surface is
  and failure may develop progressive
- Saturation of fine-grained specimer
  back-pressuring) is not possible, bu
  can be set to zero by inundating sp
- The area of the shearing surface ch
  continuously. Change of area must
  corrected for

(a)

Figure 7.10 Principles of devices to
apparatus, (c) laborator

particles and aggregations of m
than the triaxial test as it may
one, and with available shear l
give a better representation of
diameter, say, 100 mm (see Fig

Figure 7.10 illustrates the ɪ
in the laboratory, the most co
shear box. The ring shear appɛ
certain disadvantages of the cɪ
shear strength at very large sl
specimens, but can also be use

In the vane shear test, dire
soil and the surrounding matɛ
available. The vane can meas
suitably adjusting the rate of lɪ

have to be decided upon before

- minimum size of shear box
- status of consolidation and
- controlled strain or control
- rate of straining or stressing
- normal stresses to be applie
- maximum horizontal displa

Box size and shape and spec

Shear boxes are usually square (
boxes it is much easier to acco
Typical sizes for the square box
circular shear boxes common si

The maximum particle size
specimen (Cheung et al., 1988).
apply:

- the specimen thickness shou
  soil, and not less than 12.5
- the specimen diameter (or v

An alternative specification quo

- the specimen thickness shou
  size of the soil,
- the specimen size (square) c
  maximum grain size.

Cheung et al. (1988) found
samples was adequate for testing
of up to approximately 8 mm.
stress-strain curves and higher
excessive single particle load ca
the confines of the box. (It sho
20 mm thick specimens are avai

Status of consolidation, dra

Shear box specimens can be sh
solidated, undrained or drain
theoretically possible, but not
specimen drainage:

- unconsolidated, undrainec
- consolidated, undrained ((
- consolidated drained (CD

With specimens of standa
box, the drainage path is mu
pore pressures to be dissipatec
achieve undrained shear. High
because of viscosity effects. He
as only drained shear can be r

For pervious soils, the CD
strength parameters, c' and φ'.
UU or "quick" test or the CU
and preferable. Pore pressure
specimens be fully saturated. I
may be unreliable.

Tests for which drainage is
normal load and fully immerse
of capillary stresses.

Controlled strain or contro

The shear stress can be appliec
ment (i.e., stress-control), or a
stress (i.e., strain-control). Stre
if tests are to be run at a very l
kept constant) and when the cr
tests employ step-wise load ir
accurately. They are unsuitabl
decide what rate of movement

$$t_f = 12.7t_{100}$$

where t_{100} is the time to 100%
obtained by extrapolating the
consolidation phase of the test to
Equation 7.2a is based on attair
specimen. ASTM D3080 recom

$$t_f = 50t_{50}$$

where t_{50} is the time required fo
This equation gives essentially th
1988).

According to Blight (1963k
95% consolidation on the sheau

$$t_f = \frac{1.6H^2}{c_v}$$

where H is half the specimen th
When t_f has been determin
drained direct shear test can be

$$\text{Rate of shearing} < \frac{\delta_f}{t_f}$$

where δ_f is the horizontal disp
This value is not usually known

Normal loads or stresses to

The normal pressures applied to
mum stress which is likely to occ
at least four different values of
envelope.

With cohesionless soils, the
but with heavily over consolidate

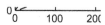

$$\begin{array}{c} 0 \\ 0 \qquad 100 \qquad 20(\end{array}$$

Figure 7.11 Effect of bonding on the
Vargas, 1974 and Rodrig

a cohesion intercept. If this co
application of the strength tes
stresses on carefully handled u
values of applied stresses are k
effect of soil bonds on the she
1974 and Rodriguez, 2005) in
Note from the test by Rodrigi
necessarily mean that the soil l

Density of compacted spec

If tests are going to be carrie(
should be defined. The angle o
of the density (or void ratio in

Maximum shear displaceme

The strain-controlled direct sh(
ing problem requires a knowle(
to carry out such tests is the
now commonly available. Mo
cycles for studying post-peak k
of any length, to reach the re:
to reach the residual strength
necessary.

 Tests that do not require
minated after the peak strengt
shear displacement. With soil:
weaker specimens) the tests sh

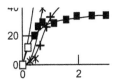

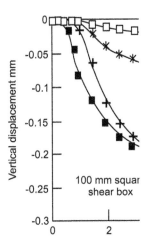

100 mm squar
shear box

Figure 7.12 Shear box tests

Examples of direct shear res

Figure 7.12 shows the results of
a soil residual from weathered
of shearing resistance with inci
(b) shows the compression or
Figure 7.13 shows 130 mm dia
shear box tests were made on

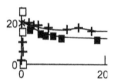

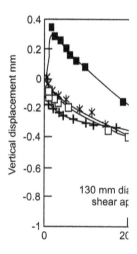

Figure 7.13 Ring shear tes

were on semi-disturbed specii
small segments of soil to form
    Whereas the shear box test
the ring shear tests were taken
sponds to a shear displacemen
stress peaked at a displacemei
ure 7.13 which obscures the p(

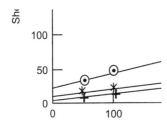

Figure 7.14 Failure envelopes for she
(a) 100 mm square shear
strength (5 reversals). (c)

100 mm square specimens, sho
reverse and forward shearing n
shear strength (a total displacen

The failure envelopes corr
Figure 7.14. In this case the r
shear strength parameters than

Figure 7.15 shows a comp
segmental ring shear specimens
by painstakingly cutting small l
into the ring shear apparatus wh
specimen. Although there were
displacement and vertical displa
residual shear strength envelope
specimens giving a slightly high

7.2.3 Triaxial test

The triaxial testing equipment
variety of test procedures to de
teristic stress ratios (e.g., K_0) o
consolidation and permeability
of cylindrical soil specimens wa

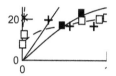

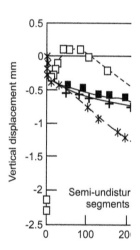

Figure 7.15 Ring shear tests on clay "undisturbed" segmenta

In practice, the following types of tests are also possible:

- unconsolidated undrained
- isotropically or anisotropi (CIU or CAU) test with or
- isotropically or anisotrop (CID of CAD) test.

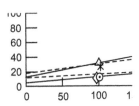

Figure 7.16 Comparison of results of i
of clay residual from sme

A diagrammatic layout of t
is sealed in a thin latex rubber n
A load is applied axially, throu
magnitude of the deviator stress
either saturated or unsaturated
internal to the triaxial pressure
In a compression test the ¿
the specimen cross-sectional ar
σ_1 {$(\sigma_1 = (\sigma_1 - \sigma_3) + \sigma_3$}. The int
equal to σ_3 and correspond to t
confining pressure equals σ_1 an
to the loading ram). The axial ¿
of the specimen enable the dra
from the top cap. Alternatively,
conditions of no drainage, or th
A standard test is usually carriec
the confining pressure σ_3 or σ_1 fc
either compression or tension. (
the specimen top cap and the l
and the loading frame. The base
frame to resist the tensile force ¿
It should be pointed out tha
tests do not necessarily match t
specimen should be representat
the specimen *in situ* and in the l
that the application of triaxial t

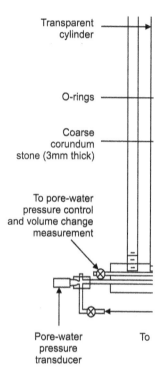

Transparent
cylinder

O-rings

Coarse
corundum
stone (3mm thick)

To pore-water
pressure control
and volume change
measurement

Pore-water
pressure
transducer

To

Figure 7.17 Triaxial c

interpretation. An evaluation
on a knowledge of how and k
details of the problem to whic

Triaxial test variables

The results that can be obtaine
available equipment are:

- the strength envelope with
- the pore pressure response

For testing residual soils, the sp
imens with smaller diameters a
effect relating to fissures and j
should not be less than 8 times

The ratio of specimen lengt
than 3 to 1 to minimize both e
compressive stress.

Consolidation prior to shear

The specimen is either consolid
shear, or no consolidation is all
respectively). In saturated soils
same depth, the compressive str
found to be independent of the
and compact silts at low cell pre
is approximately horizontal, i.e.
cohesion c_{uu}:

$$c_{uu} = \frac{1}{2}(\sigma_1 - \sigma_3)_{max}$$

where $(\sigma_1 - \sigma_3)_{max}$ is the devia
there is a unique relationship
unconsolidated undrained (UU)

For residual soils, which ar
strength will increase with incre
be a straight line, but is often cu
air in the voids becomes compre
be large enough to cause full sat
a set of UU test results for an un
UU strength envelope in terms
stress envelope is a straight line

Consolidation stress system

The confining stress system duri
or anisotropic ($\sigma_2 = \sigma_3$ and $\sigma_1 >$

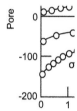

Changes in pore w

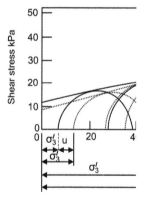

Mohr circles at fail

A - Total stres
(φ approa
B - Effective

Figure 7.18 Typical results of UU tri:
stress has been taken a:

$(\sigma_1 - \sigma_3)$ is usually applied th
stress conditions in the groun
an anisotropic stress system ¿
than isotropic consolidation.]
with $\sigma'_3 = K_0\sigma'_1$, so-called K_0-c

pore pressure is subtracted from
stress paths are:

- compression loading   (σ
- compression unloading   (σ
- extension loading   (σ
- extension unloading   (σ

Other stress paths sometimes er

- constant mean principal str
- constant stress ratio
  where q′ = ½(σ′₁ − σ′₃).

Undrained strength, effectiv
between compression and exten
The highest values are usually o
As examples of loading an
compression and extension tests
residual from a micaceous schis
and φ′ = 39° and 31° for the com
illustrated in Figure 7.20, show
in the soil. In extension, failur
compression the failure surfaces
soil was unsaturated and u_a an
In each case, the strength lines
saturated soil and therefore rep
Bishop & Wesley (1975) de
capable of any stress path test.
only been designed for specime
cell can be used for stress path
supported by a hanger on the loa
is modified to apply a controlle
actuator attached to the reactior
to be practical (Baldi et al., 198

90° 0

ⓐ

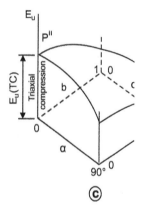

ⓒ

Figure 7.19 Hypothetical surfaces re
et al., 1988): (a) undraine
σ₁ and σ₃ vary in a singl
P and Q of the surface.

Saturation conditions and I
(for CU and CD tests)

Residual soils are usually unsa
is considered routine in many
    Saturating a residual soil
than the natural state usually e
is often deep and foundation:
may happen that the ground v
represents the least favourable
little effect on the friction angl
in cases where saturation cause
will the cohesion in terms of e

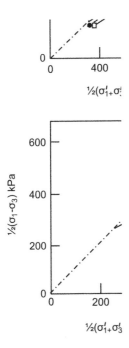

$\frac{1}{2}(\sigma'_1 + \sigma'_3)$

$\frac{1}{2}(\sigma_1 - \sigma_3)$ kPa

600

400

200

0

0   200

$\frac{1}{2}(\sigma'_1 + \sigma'_3)$

Figure 7.20 Comparison of results c
residual micaceous schis
unsaturated specimens (σ
laminations (b) σ_1 paralle

The extent to which satura
checked by measuring Skempto
eter is, however, dependent on t
of both soil structure and pore
same B-value, the degree of satur
Higher saturation will exist in t

Saturation by applying a ba
of the specimen and thereby cor
to Boyle's and Henry's laws resp
theoretical back pressure Δu_{bp}

$$\Delta u_{bp100} = 49 u_{a0}(1 - S_0) \; v$$

To saturate from $S_0 = 70\%$, the
1470 kPa. (Also see Section 4.

Single-stage and multi-stag

In conventional triaxial testin
system, a new specimen is set
series should be as near identic
most residual soils because of
the shear strength parameters,
When failure has started to de
released and the lateral stress
the deviator stress is again ine
confining pressure. This proce
three) have been obtained or
reliable results to be expected
the multi-stage test can save tes
(However it is doubtful if sm
quality of the test results can b
 Lumb (1964) applied the
urated residual soils from He
indistinguishable from the sing
necessary to obtain at least thr
turbed specimens. The largest
contains relict bonds such as th
not be used as the effect of fai
cannot be determined.
 Ho & Fredlund (1982b) de
soils to measure the increase in
tities c', φ' and φ^b can be obtai
out in Figure 7.22. $(\sigma_3 - u_a)$ is
increased by increasing u_a.

Figure 7.21 Principle of multi

Controlled strain or controll

The most convenient way to
axial strain. The controlled-st
loading. Lundgren *et al.* (1968
stress-controlled shearing:

Advantages:

- Load increments may be sele
  pore pressure equalization i
- The deformation versus ti
  increment.
- For structurally sensitive so

Disadvantages:

- Failure may be abrupt and
  nation of the ultimate stren
  easily possible.
- In drained tests, applicatior
  in failure) will usually cau
  conditions.
- In undrained tests, the pore
  be measured accurately.

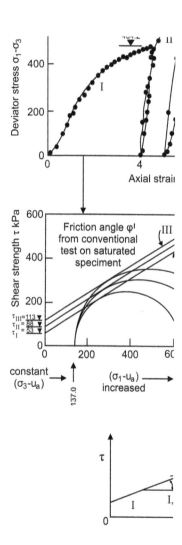

Figure 7.22 (a) Determination of th
method and multi-stage
III is reduced due to dis

several hundreds of kPa (Bishop

The method used to measur
pored filter (normally of ceramic
(see Figure 7.17). The pore water
required to prevent flow of wat
filter must have a sufficiently hi
displacing water from the pores
pressure required to displace w
the air entry pressure of the porc
enter the porous stone and wate
measured on the remote side of
then be the pore air pressure a
minimum air entry pressure of p
triaxial tests is 500 kPa.

Cell and consolidation press

Cell pressures in CU and CD te:
With a saturating back pressure
effective consolidation pressures
of shearing resistance is almost i
was originally unsaturated. Als
hand, high stresses in combina
of weak bonds and/or particle c
resistance. At low stresses, the st
by using consolidation stresses
may be of engineering importanc
where the overburden stress ma

The consolidation stresses a
20, 40, 60 and 80 kPa). Four tes
adequately.

Rate of strain

In undrained tests without pore
c_{uu} decreases with increasing ti
Wilson, 1951). For clays, the d

0

Figure 7.23 Typical pore pressure c
during rapid shearing (E

to failure. A commonly used t
strain per minute.

For undrained tests (UU ɩ
deformation must be slow enou
to equalize and in drained test
This results in similar times to

In undrained tests the non
from the non-uniformity in str
pore pressure at the ends is us
when measuring the pore pres
is too fast for equalization, the
the position of the failure env
(Bishop *et al.*, 1960). The tim
and the coefficient of consolid

Figure 7.23 shows pore pre
and centre (u_c) of two triaxial
the normally and overconsolic
of the specimen was greater th
stress would have apparently k

Figure 7.24 shows experi
two clays which were run at ɑ
those calculated from the theoɩ
indicate that actual equalizatiɩ
theory suggests. This is despite
drains (and in particular, papɩ
not.

It is apparent from Figure
ization the error in the value oɟ
stress history of the soil. With
the value of σ'_3 may occur even
With normally consolidated so

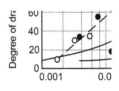

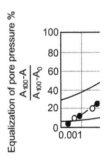

Figure 7.24 (a) The relation between
relation between equaliza
1963b).

even with testing rates which a
In deciding on a test duration f
of equalization must be conside
degree of equalization. Lower d
are normally or lightly overcoi
aimed at in tests on heavily ov
along a narrow zone).

For tests without drains:

$$t_f = \frac{1.6H^2}{c_v}$$

where 1.6 is the time factor corr
and H is half the specimen heig

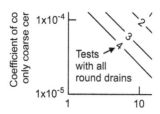

Figure 7.25 Chart for finding test

Similarly, for test with all-

$$t_f = \frac{0.07H^2}{c_v}$$

The meaning of the term '
the object is to measure the pe
of the test may be taken as th
on the whole stress path is req
of the test and the first signific
In Figure 7.25 equations 6
which enables a test duration
knowing c_v, the coefficient of c

Triaxial testing of stiff fissu

Clays formed by the *in situ* w
relict joints that are commonl
izontal. Because of the unfavc
triaxial tests on these soils ofte
planes become visible at an axi
be confined to this surface witl
failure surface acting as relativ
falls after the appearance of th
plane is continually decreasing

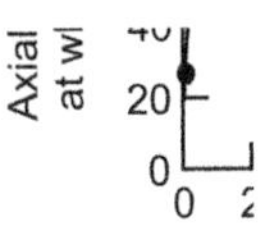

Axial
at wl
20
0
0

Figure 7.26 Stress-strain curves for

Figure 7.26 shows the stres
stiff fissured clay. The specimens
a thickness of 0.15 mm and we
0.05% per hour. The effective c
rams of the triaxial cells were g
of ram friction could be assesse

In Figure 7.26 the axial str
been allocated a relative value
after the formation of the failur
stationary (S).

It appears to be generally
(1960), Bishop *et al.* (1965)) th
friction. If this is the case, it wil
curves with the bushes stationar
of 5 to 10% ram friction accoun
The remaining increase in devi
ascribed to restraint developed l
ends of the sliding blocks of soi

Bishop & Henkel (1962) c
exceed 14 kPa at axial strains o
the problem shows that, if the m
local strains in the rubber may l
soil specimen. This together wit
corrections for stiff fissured clay

An investigation into memb
by 76 mm high rigid dummy sp

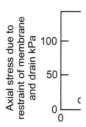

Axial stress due to
restraint of membrane
and drain kPa

100

50

0

0

Figure 7.27 Membrane and drain r
(a) Dummy specimen u
dummy specimens.

Figure 7.27a) was cut along a c
The two faces of the pre-cut "f
cylinder was roughened using
silicone grease, the dummy spe
cell with a rotating bush. The
was only 0.02 (Blight, 1963b)
almost entirely due to membra

Figure 7.27 (derived from
indurated clay specimens) sho
stress and the increase in axia
which a shear plane first bec
specimens represents the effec
compared with results from t
this comparison that the dumr
behaviour of specimens of ind

Selected measurements fro

The axial strain referred to
of a failure surface. To apply
strain, this failure strain must
watch the triaxial specimen c
surface first appears. This stre

Examples of triaxial test res

Figures 7.28 to 7.30 are examp

    Figure 7.28 shows typical c
residual andesite lava. Figure 7.
for the soil with, inset, values of
shows failure stress points for sp
ent sites within a distance of ab
variability of residual soils, all
a number of different lava flow
vary (Blight, 1996).

    Figure 7.29 (Bishop & Blig
made in consolidated undraine
rated soil at increasing values o
have been measured separately i
ure 7.17. It will be noted that as
Also that in each test $(u_a - u_w)$
then increased as the soil started
the soil became saturated by con
urated as the soil dilated at larg
an increase in volumetric strain.

    Figure 7.30 (Bishop & Blig
those shown diagrammatically i
equalization of pore pressure e
of u_a at the base and mid-heigh
of the strength line for saturated
and u_w for the unsaturated soil.

7.2.4 Determination of I

The coefficient of earth pressure

$$K_0 = \frac{\sigma'_h}{\sigma'_v} \quad \text{for } \varepsilon_h = 0$$

where the subscripts v and h refe

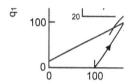

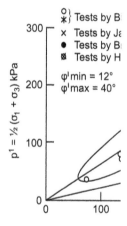

$p^1 = \frac{1}{2}(\sigma_1 + \sigma_3)$ kPa

$\left.\begin{matrix} O \\ * \end{matrix}\right\}$ Tests by B

× Tests by Ja

● Tests by B

▨ Tests by H

$\varphi'min = 12°$
$\varphi'max = 40°$

Figure 7.28 (a) Typical stress paths f
(b) Summary of triaxial
five different sites in the
measured pore pressur

The usual object of K_0 te
for both normally consolidat
have been used to evaluate K_0

(i)   For zero lateral yield to
vertical compression mu:
by the vertical compress

Figure 7.29 Pore pressure changes
residual from weathered

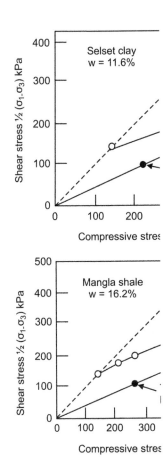

Selset clay
w = 11.6%

Shear stress ½ (σ₁-σ₃) kPa

Compressive stres

Mangla shale
w = 16.2%

Figure 7.30 Triaxial shear at constant
saturated soils. Degree
shale 90%.

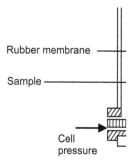

Rubber membrane

Sample

Cell
pressure

Figure 7.31 Triaxial test with a

Starting with a fully c
at a rate slow enough to
σ'_h is adjusted continuou
This is a very simple
appears to give reasonal
(ii)   The horizontal strain ε_h
tinually adjusting σ'_h as
horizontal strains. The n
cator (Bishop & Henkel
of a brass half-loop, or
with a LVDT (linear vol·

Figure 7.32 shows a set
smectitic mud-rock. (See also F
relationship between the applie
lateral strain. Note that the re
has a value of about 0.63. As t
a value of 1 when σ'_v reaches 4
Figure 7.32b shows the
during initial consolidation, th
1.47 (6%). Hence the soil wa
the test.

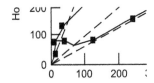

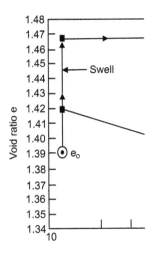

Figure 7.32 K₀ measurements made

K₀ can also be measured ir
for a residual andesite lava prof

7.3   FIELD STRENGTH TE

7.3.1   General

Field tests are advantageous for

- The disturbance caused by s
  mostly eliminated.

Cone penetration test	For
	Sem
*Pressuremeter (prebored, or self-bored)	For
	Diffi
*Plate bearing test or screw plate bearing test	Usu:
	evalɪ
	inteɪ

*See sections 6.4.1 to 6.4.4 for details.

- The test specimen size car
  soil mass. There are many ɪ
  or by an indirect measure.
  found wide-spread use in
  lists some of these tests an
- Most residual soils behave
  by shearing are therefore ɪ
  effect can thus usually be ɪ

When planning a site inv
shear strength parameters, on
These have to be used in an ɪ
and time, and in combination
of the available methods mus
actually needed for the design ɪ
influence the design. It is theref
understanding of the various ɪ
the test variables, and of the
investigation program and late
   In this section, the proced
borehole shear, the Standard I
(CPT) will be described. Pres
measure stress-deformation pɪ
to failure (which may be diffic
calculate strength parameters.
terms of compressibility and sɪ
do not measure the strength dɪ

The main purpose of the tes
for either the intact material or
test is generally carried out at the
in shafts.

Most of the tests are set u
the shear plane should be paral
slickensides) or coincide with a
surface of a slip).

The size of the specimen sho
specimen sizes are 300 × 300 m

Excavation of the test pit a
with utmost care to avoid distur
the specimen, handsawing and c
shaped, it must be protected usin
The final trimming must be don
content. Special precautions are
effects of water pressure and se
specified discontinuity, the spati
identified in terms of strike dire

The equipment for applyin
hydraulic jacks, flat jacks actin
anchor system. It is important tl
normal load to the sample and n
test the alignment of the norma
increases.

The system for applying the
of shearing. Reactions are often
a bulldozer. In certain cases, a b
load. It is important to allow f
deformation measurement syste
to reset the deformation gauges
available with longer travels and
Applied loads should be measur
value. Movement reference poir
test to ensure they are not inflt
variations on displacement shc

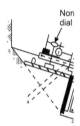

Nori
dial

Centralizing fram

Figure 7.33 Stages in performing an
(b) Cutting sides of soil

the test set-up and correcting
monitor temperatures at sever:

Examples of in situ direct s

Field direct shear tests carried
by James (1969), Mirata (197‹
(1988), and others with variou
 James (1969) reported on
dam site. The test blocks were
anchor system with the anchor
normal pressure of 600 to 80(
against a concrete block built
0.5 mm/min. The test was carr
load, the load was released wh
was applied and time allocate
the block was restored to its o:
was then repeated under a nor
reached.
 Mirata (1974) introduced
wedge of soil is sheared along i
The test has been applied in u

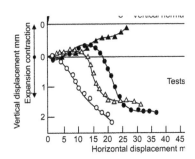

Figure 7.34 Results of field direct she
apparatus: (a) Shear stre:
residual).

stability problems. Its principle
respect to the direction of load:
stress can be varied over approx
in slope stability problems. The
in conjunction with a total stre:
applicable to saturated soils abc

The field shear test proced
for use in residual soil slopes.
305×305 mm and 150 mm dee
tion from a light steel frame lc
means of a hand-driven screw ja
by modifying a field CBR appa
controlled hydraulic jack was en
socket bearings. The normal str

Typical test results obtainee
tests were carried out at two lo
ferent grades of weathering. Lo
granite), while location B was i
location A two test series were
the other soaked (A-2). Soakin
tic sheeting and submerging the
water for about 12 hours. At
content.

water content amounted to 13

7.3.3 Vane shear test

The test is usually used only
penetration and rotation of tl
comprises low strength clayey
and pedogenic nodules. Howe·
ual soils with peak strengths
that the vane shear test is not
This is well illustrated by the r
shear strengths far exceeded th·
back-calculated from a slide fa
vane shear strength approache
are to clays and silts originati
which do not contain gravelly

Principle of the vane test

The vane usually consists of fou
at right angles to a torque rod i
A torque is applied to the rod l
causing the blades to rotate a
surface. The shear strength is c
to shear the clay along the cyli

The vane shape commonl
width ratio, H/D = 2. The aci
soil to be tested. For example,
for strengths between about 5(
smallest size suitable for accu
However, smaller vanes (as sn
residual soils (e.g., Blight *et al.*

Vane tests may be carried
or by pushing the vane into tl
vane can also be pushed into
ground surface is rarely possi

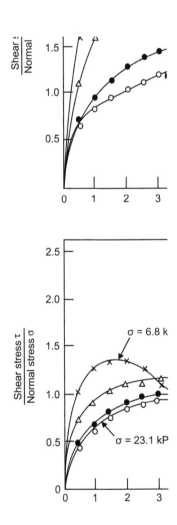

Figure 7.35 Normalized stress-displac
Office direct shear mach
soaked conditions (Brand

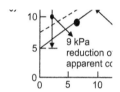

10

9 kPa
reduction o
5   apparent c(

0
0   5   10

Figure 7.36 Strength envelopes for
envelope obtained from
cohesion caused by soa
material and with the d

high. Figure 7.37 is a diagram
soils.

Figure 7.5 showed the resu
Figure 7.38 shows a vane stre
that in this profile, as with tha
agrees reasonably well with th
triaxial tests, whereas the mea
the effects of soil discontinuiti
strength of the intact soil and
material.

Effect of vane insertion

Drilling a hole for the vane tes
of the hole. Tests at the Norv
indicated that a vane should
six times the borehole diamete
diameter hole. However, in stif
to be a lot less, and a distance

Mode of failure

Finite element analyses have i
tical sides of the cylindrical fa

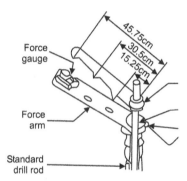

Force gauge

Force arm

Standard drill rod

45.75cm
30.5cm
15.25cm

Alternative torque application
and measuring arrangement

Figure 7.37

assumption of being uniform (V
cylinder, the shear stress distri
edges. However, with H/D = 2
arises from the cylindrical surfac
14%. The failure mode is mos
whereas the direct shear mode
failure surface is formed (Chand
ual soils has been checked by sa
sampler and exposing the failur
and very little disturbance is vis

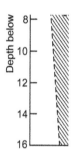

Figure 7.38 In s

Shearing under undrained ɛ

In order to enable a definitive
place either under completely ι
conditions can be assured if, fo
in the failure zone U is less tha
lished by Blight (1968) based
data. It takes the form of an e:
the time factor

$$T = \frac{c_v t_f}{D^2}$$

is related to U (see Figure 7.39
where c_v is the coefficient of c
   t_f the time to failure,
   D the vane diameter.

For a commonly used vane size
U vs T drainage curve, T is less
about $110 \, \text{m}^2/\text{y}$ or 3.5×10^{-2} ɛ
usually satisfied with all claye
sand lens. Obviously, very mu
strength is to be measured. In t
This type of very slow vane tɛ

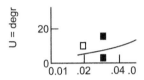

Figure 7.39 Empi

a motorized slow drive to rota
In silty soils, it is possible to n
shearing time. U will exceed 9(
least 6 minutes.

Vane size and shape

The vane can be of any size, but ι
minimize disturbance effects on
The area ratio is defined as the
2Dt with t = blade thickness) to
or 8t/πD.

As stated above, smaller va
ple, Blight (1967a) reports on a
salts with an undisturbed vane s
apparatus had blades with H =
to an area ratio of 17%. As vane
carried out previously, the effec
gated. A series of tests was carri
clay derived from the *in situ* wea
tests ranged from 5 seconds to
very slight decrease in measure
decrease, however, was less than
time to failure. As a result of thi
has been adopted for tests on in

In addition to the standard
with other ratios and also triai

Figure 7.40 Effect of time to fai

angles have been employed,
(Aas, 1965; Blight *et al.*, 197(
Silvestri *et al.*, 1993).

For calculating the shear s
strength is fully mobilized anc
The expression for the vane to

for a rectangular vane: T :

for a triangular vane: T :

where S_v and S_H are the undr.
planes respectively and S_β is tl
the horizontal, and L is the len;
of vane tests using vanes of v.
weathered mudstone (Blight *et*
larger than S_v. This arose bec:
which horizontal stresses paral
by the slide. Figure 7.41b show
from sandstone which showed
If $S_v = S_H = S$, equation 7.

$$T = S\pi D^2 (H/2 + D/6), \ ($$

$$S = 2T/\{\pi D^2 (H + D/3)\}$$

The horizontal to vertical :
measured by means of vanes h

$$S_v = KS_H,$$

so that equation 7.7a becomes

$$T = \pi D^2 (KHS_H/2 + DS_H$$

$$K = 2T/\{\pi D^2 S_H (H + D/:$$

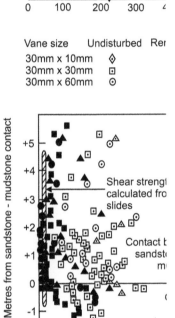

0   100   200   300   ⌐

Vane size   Undisturbed   Rer

30mm x 10mm   ◊
30mm x 30mm   ⊡
30mm x 60mm   ⊙

Metres from sandstone - mudstone contact

+5

+4

Shear strengt
calculated fro
slides

+3

+2

Contact t
sandst
m

+1

0

(

-1

0   100   200

Strength kPa

Figure 7.41 (a) Directional strength
shapes of vanes in a resid

Remoulded vane shear stren

Once the undisturbed or peak sh
to 25 times and the torque is rer
remoulded shear strength. As n
and 7.42, the remoulded stren
or joint in the soil. In stiff joint
approximates to the strength of

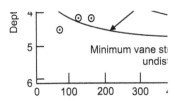

Figure 7.42 Laborator

Comparison of vane shear
other types of measuremen

Figure 7.5 shows a comparisc
from weathered shale with unc
on 76 mm diameter specimens
a 76 mm diameter circular sh
tests correlate quite well with t
similar comparison for a soil re
Here again the remoulded van
scale laboratory shear tests. I
shear strengths are somewhat
back-calculated from large-sca
results do not differ very muc
indicative of the low A value c
of Figure 7.42 shows that repe
with the strength calculated fc
mass is controlled by the streng

7.3.4 The pressuremet

The use of the pressuremeter t
been referred to in Chapter 6
test can be used as a measure c
the relationship that is used to

$$c_{uu} = \frac{p_L - \sigma_{vo}}{N_c}$$

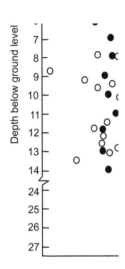

Figure 7.43 Comparison of strength:
pressuremeter with $N_c =$

If the pressuremeter is carr
prevail, a drained strength c_D ca

$$c_D = \frac{p_L - \sigma'_{vo}}{N'_c}$$

Pavlakis (1983) has shown
pressuremeter results and uncon
His results, for a very soft rock r
Figure 7.44 shows the pressu:
strengths. As many of the tests d
indicate that strengths should b
ial shear strength envelope of F
validity, provided the correct va

7.3.5 Standard penetrat

The appropriateness of the SPT f
stated that the test may at most

Figure 7.44 Pressuremeter data of profile.

This critique may have its just
the test has a poor reproduci
demonstrated that under labor
within a standard deviation o
however, not only due to the
also because of the variety of
lack of enforcement of equip
probably continue to be used
of its shortcomings, because i
conditions in both difficult an

Principles of the test

Present practice involves drivin
construction (see Figure 7.45)
(or soft rock) at the bottom of
of a 63 kg hammer with a free
tube the first 150 mm are con
at the bottom of the hole. The

Figure 7.45 Stan

next 300 mm is termed the SPT ι
withdrawn, dismantled, and the

Split spoon sample tube

The thick-walled split spoon ɦ
457 mm. There is a driving sh
rel, and a coupling at the other ε
valve to prevent sample loss. Sσ
prevent loss of sample.

Australian, British and Souι
cone to replace the open shoe wh
to the cutting edge of the drive s
(1957). The N-values are of siι
When applied in loose and med
the cone may give significantly ι
The SPT was originally devι
ported soils. As most residual s
the SPT have had to be made. A
from tests on various stiff clays
the ratio c_u/N where c_u is the unς
index. For plasticity indices betν
5 kPa. The ratio appears to be
spacing. SPTs in clays may be r
turbed samples by means of thι
out that in fissured clays the m
strength of the intact material.
applicable to clayey residual so
correlation provides a first appr
and is taken as:

$$c_u = 5N \text{ [kPa]}$$

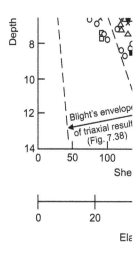

Figure 7.46a Variation of SPT "N

The SPT N value has also beer
soils by the equation

$$E = 200c_u = 1000N \text{ [kPa}$$

Figure 7.46a shows the v
on residual andesite lava. As i
strength is reasonably good at c
expression appears to overesti
for the same weathered andesi
much greater depths (50 m as c
depth (see Figure 7.46b). This
represented in Figure 7.46b wa
This illustrates one of the man
applying empirical relationshij

7.3.6 Cone penetration

The quasi-static cone penetrati
ited extent. Residual soils are c

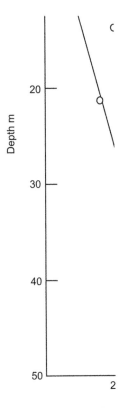

Figure 7.46b Variation of SPT "N" with

to the top few metres. Moreover,
ules, which often occur in residu
cone. Still, the survey by Brand
is fairly widely used in residual

Field penetrometer testing o

As the modified vane apparatus
600 kPa, soils with an undistur

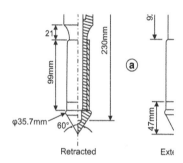

21

99mm

230mm

φ35.7mm 60°

Retracted

9

47mm

Ext

(a)

Figure 7.47 Two common types of sleeve cone.

by an alternative means. One
the cone penetration test. The
Delft mantle cone and the Beg
of slightly reduced diameter a
downward movement of the c
the cone (penetration resistanc
applying a thrust on the inner
second measurement is taken
and friction sleeve are pushed

In the electrical penetrome
force on the friction jacket are
into the cone. Electric cables t
(e.g., solid state memory) trar
The electrical cone penetrome
resistance, as well as pore wate

Electrical penetrometers ai
measurements with higher pre
they allow the simultaneous 1
sleeve friction. There is the op
verticality of the sounding. Th
it is pushed down, resulting i
exceeding a certain depth or
corestone.

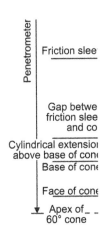

Penetrometer

Friction slee

Gap betwe
friction slee
and co

Cylindrical extensio
above base of con

Base of con

Face of con

Apex of_ _
60° cone

Figure 7.4

The piezocone is a more r
pressure transducer within the t
this purpose should have a volu
Figure 7.48 illustrates the Fugrc
of the cone penetrometer in resi

The relationship between th
resistance q_c is of the form:

$$c_u = \frac{q_c - \sigma_{vo}}{N_c}$$

where σ_{vo} is the total overburde
N_c is a bearing capacity f

σ_{vo} is usually negligible in c
7.12 with little error. Penetrom
the bottom of a 100 mm diame
its closed position to a distance
then advanced separately until
of the effects of the rate of pen
resistance is usually reached af

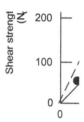

Figure 7.49 Correlation between cc
for two indurated clays.

25 mm per minute has arbitrar
to failure in the penetrometer
soil-specific.

To evaluate N_c for stiff res
were made by Blight (1967a) at
Strengths of up to 280 kPa we
clay, while strengths over 300
The results of the comparativ
penetration resistance to undi
N_c appears to be 15.5. In most
is about twice the unconsolida
resistance to triaxial shear str
agrees fairly well with extreme
and Ward *et al.* (1965) for cor

Marsland & Quarterman
cone penetration data with s
obtained from various stiff cl
a trend of N_c to increase with
distinct influence of the scale (
in the clay in relation to the c
distinguished, as shown in Fig
of discontinuities, N_c values as
discontinuities, $N_c = 15$ woulc

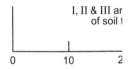
I, II & III ar
of soil f

0   10   2

Close spacing I

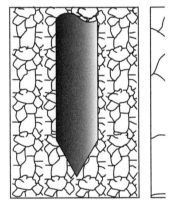

Figure 7.50 (a) Cone factors for stiff
fabric features with relati

For normally- and over-con:
shear strengths via a pore press

$$B_q = \frac{u_d - u_e}{q'_c}$$

where u_d = dynamic pore wate
$\quad\quad u_e$ = equilibrium pore wa

both measured by means of a p
The relationship between B
experimental points are widely :

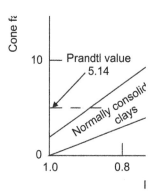

Figure 7.51 Summarized relationsl
$B_q = (u_d - u_e)/q'_c$.

shown in Figure 7.51, but the
bearing capacity value (Prandt

$$N_c = \pi + 2 = 5.14$$

for normally consolidated cla
which agrees with the $N_c = 30$

REFERENCES

Aas, G. (1965) A study of the effe
  in situ shear strength of clays. (
  Vol. 1, pp. 141–145.
Andresen, A. & Simons, N.E. (19
  Conf. Shear Strength of Cohesi
Andresen, A. (1981) Exploration.
  Brenner, R.P. (eds) *Soft Clay En*
Baldi, G., Hight, D.W. & Thom
  methods. In: *Advanced Triaxi*
  USA, ASTM. pp. 219–263.

Bishop, A.W., Green, G.E., Garga, ⌐
apparatus and its application to ⌐
273–328.

Black, D.K. & Lee, K.L. (1973) Satι
Found. Div., ASCE, 99 (SM1), 7

Blight, G.E. (1963a) Bearing capaci
Civ. Engrs'. Diamond Jubilee Co

Blight, G.E. (1963b) The effect of ⌐
the shear strength of soils. In: *Lab*
USA, ASTM.

Blight, G.E. (1963c) Effective stres
soil. *3rd African Reg. Conf. Soil l*

Blight, G.E. (1967a) Observations
Conf. Shear Strength Properties ⌐

Blight, G.E. (1967b) Effective stres
Div., ASCE. 93 (SM2), 125–148

Blight, G.E. (1968) A note on fiel
142–149.

Blight, G.E. (1969) Foundation fai
Div., ASCE, 95 (SM3), 743–767

Blight, G.E. (1984) Uplift forces m
Soils, Adelaide, Australia. pp. 36

Blight, G.E. (1985) Residual soils
Sampling and Testing of Residu
Int. Soc. for Soil Mech. & Founc

Blight, G.E. (1996) Properties of a
Soils, Kuala Lumpur, Malaysia. `

Blight, G.E., Brackley, I.J. & vaι
Bethlehem – an examination of t
June, pp. 129–140.

Brand, E.W. (1988) Some aspects
Symp. Field Meas. in Geomech.,

Brand, E.W. & Phillipson, H.B. (1
South East Asian Geotech. Soc.,

Brand, E.W., Phillipson, H.B., Borr
Kong residual soil. *Int. Symp.*
pp. 13–17.

Brenner, R.P., Nutalaya, P. & Berι
properties of granite residual so
Madrid, Spain. Section 2, Vol. 1,

decomposed granite. *2nd Int. C*

Chu, B.L., Hsu, T.W. & Lai, (198
    Int. Conf. on Geomech. in Tro

Cowland, J.W. and Carbray, A.N
    saprolitic soils. *3rd Int. Conf. (*

De Ruiter, J. (1982) The static co
    Amsterdam, Netherlands. 2, 38

Florkiewicz, A. & Mroz, Z. (198
    on Soil Mech. & Found. Eng.,

Fredlund, D.G., Morgenstern, N.
    soils. *Canadian Geotech. J.,* 15

Garga, V.K. (1988) Effect of sam
    Geotech. J., 25, 478–487

Gibson, R.E. & Henkel, D.J. (19
    measured 'drained' strength. G

Head, K.H. (1982) *Manual of Sc
    and Compressibility Tests.* Lon

Ho, D.Y.F. & Fredlund, D.G. (19
    soils. *ASCE Conf. Eng. and*
    pp. 263–295.

Ho, D.Y.F. and Fredlund, D.G. (1
    Geotech. Test. J., 5 (1), 18–25.

Howatt, M.D. (1988) The *in sit*
    Geomech. in Tropical Soils, Sin

Howatt, M.D. (1988) Written disc
    Vol. 2, p. 603.

Howatt, M.D. & Cater, R.W. (1
    Int. Conf. Geomech. in Trop
    pp. 371–379.

Irfan, T.Y. & Woods, N.W. (1988
    Int. Conf. Geomech. in Tropica

James, P.M. (1969) *In situ* shear te
    London, UK. pp. 75–81.

Khalili, N. & Khabbaz, M.H. (19
    strength of unsaturated soils. G

Lo, K.Y. (1970). The operationa
    pp. 57–74.

Lo, K.W., Leung, C.F., Hayata
    the weathered Jurong. *2nd I*
    pp. 277–285.

Marsland, A. & Quarterman, R.S.
tation of quasi-static penetratio
Amsterdam, Netherlands. Vol. 2,

Mirata, T. (1974). The *in situ* wedg
24, 311–332.

Palmer, D.J. & Stuart, J.G. (1957)
correlation of the test *in situ* wit
Eng., London. Vol. 1, pp. 231–2

Pavlakis, M. (1983) Prediction of F
Tests. PhD. Thesis, Witwatersran

Peuchen, J., Plasman, S.J. & van St
12th East Asian Geotech. Conf.,

Powell, J.J.M. & Quarterman, R.S.
with particular reference to rate e
Vol. 2, pp. 903–909.

Prandtl, L. (1921) Uber die Einc
Festigkeit von Schneiden. *Zeits*
pp. 83–85.

Premchitz, J., Gray, I. & Massey,
founded in weathered granite. *2r*
pp. 317–324.

Richardson, A.M., Brand, E.W. &
soft clay. *ASCE, Spec. Conf. In*
pp. 336–349.

Rodriguez, T.T. (2005) Colluvium
Portuguese. Quoted by Futai, M
tory behaviour of a residual trop
Characterization and Engineerin
pp. 2477–2505.

Serota, S. & Lowther, G. (1973) SP

Silvestri, V. & Aubertin, M. (1988)
Soils: Field and Laboratory Studi

Silvestri, V., Aubertin, M. & Chap
various vanes. *Geotech. Test. J.*,

Sowers, G.F. (1985) Residual soils
(eds) Sampling and Testing of Re
Scorpion. pp. 183–191.

Stroud, M.A. (1974) The Standarc
Euro. Symp. on Penetration Test

Thomas, D. (1965) Static penetrati

Chapters 1 to 7 have described t
behaviour and properties differ f
residual soils and the difficulties
been set out, in addition to met
by a description of site explorati
for residual soils.

Because residual soils very
mates where the wet and dry seas
a seasonally or permanent state
the concepts and methods of u
and integrated into the descript

Two more technical aspect
paction and the mechanics of (
flow through saturated and un
chapters, on compressibility, ser
and the measurement of shear s

Because Geotechnical Engi
and the field, most of the imp
the development of new enginee
from the need to solve practical
It is therefore appropriate to en
illustrate applications of the tec
geotechnical problems. In orde
following:

- settlement of two tower blc
- settlement of an earth dam
- settlement of an apartment
- pre- and post-heaving of ex
  for damage by differential H
- prediction of the rate of hea
  expansive clays,
- design investigation for pil
  siltstone and alluvial clay,

Hong Kong, India, Nigeria, S
foundation are also widely use

Developments in the field
generally overshadow more re
remains as relevant today as it

Two classic settlement st
andesite lava in Johannesburg
(1978) and Pavlakis (1983). Be
profile described in Figure 1.
method to predict the settlem

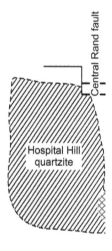

Figure 8.1 **Section through the Tot**
(Jaros, 1978).

In the case of Total House, Jar(
for Guardian Liberty centre, hi
the discrepancy in the case of '
sions or floaters (see Figure 8.1
be considered in the analysis,
certainty.

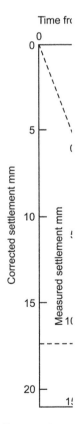

Figure 8.2 Time-settleme

~~cons can be predicted with acc~~

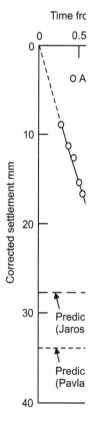

Figure 8.3 Time-settlement cur

expelled pore water reduces, aii
expelled in preference to the w
flow rather than water flow anc
and 5.8.)

If the instantaneous compre:
in Figure 8.5, it will usually l
the major part of the total con
behaviour observed for Total
Figure 8.3), both structures bei
settlement of an earth dam, the
settlement will usually be more
occurs during construction and

The data for Figures 8.4 aı
dam, constructed in 1957/58 i

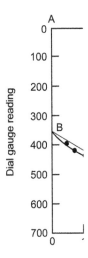

Figure 8.4 Typical oedometer time-s
measured during design of

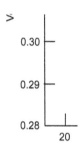

0.30

0.29

0.28

20

Figure 8.5 Relationship between voi
measured during design c

the data given in Figures 4.9 a
and later calculated by Anna
design the dam and monitor i
(1970). Figure 8.6 shows deta
The settlement calculation co
settlement beacon located at th
foundation.

The assumption was mad
air-filled voids in the compact
sure, as shown in Figures 8.4
and dissipated during constru
pressure was considered by m

$$\frac{\delta u_a}{\delta t} = c_v \left(\frac{\delta^2 u_a}{\delta x^2} + \frac{\delta^2 u_a}{\delta y^2} \right) -$$

In equation 8.1 the increa
increased was represented by:

$$\frac{\delta u_a}{\delta t} = B_a \cdot \gamma \cdot \frac{\delta h}{\delta t}$$

where $B_a = u_a / \gamma h$ and B_a is less

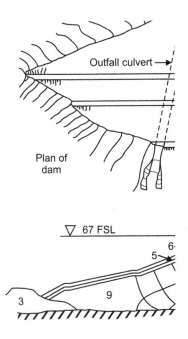

Outfall culvert →

Plan of
dam

▽ 67 FSL

6·

5→

9

3

0 10   30   5

Sc

Figure 8.6 Plan and typical section of l
and indicating steady-seep;

The remaining term repres
atmospheric pore air pressure.
conventional means, as indicate

0.7
0.8
0.9
1958

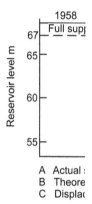

1958
Full supp

Reservoir level m

67
65

60

55

A Actual
B Theore
C Displac

Figure 8.7 Post-construction time-s
as record of reservoir w

The upper diagram in Fig
tlement calculation (lines B f
settlement (line Oabcdefgh) ar
The lower diagram in Figı
to 1962. Comparing this diag
diagram, it is clear that much o
taneous response to the incre
diagram corresponds to the inc
level diagram is only slightly a
slightly below b. Level f' is o
occurred between points d an
significant settlement occurrec
once this had been reached, re

Post-construction settlemen

The actual post-constructic
only 9% less than the measured

8.3 SETTLEMENT OF AN
IN BELGRADE

A 13-storey apartment building
1.5 km distant from the right bar
on strip footings with an aver
showed that the thickness of the
Settlement analyses predict
settlements up to 30 mm. Loess
ically disturbed. Accordingly, d
and used in the settlement analy
settlement values were several tii
gations showed that the addition
of the loess beneath the founda
the measured settlement versus
the end of 1973. It is quite clea:
until approximately half of the l
settlement occurred. This had v
settlement occurred in May–Jur
end of 1973, but it is quite lik
example, during an exceptional

8.4 PREHEAVING OF EX

A large-scale field experiment, r
essential requirements for effer
the mechanism of preheaving.
erate the penetration of water :
maximum flow path. In the exp

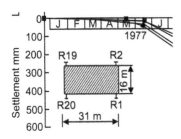

Figure 8.8 Settlement ı

holes to facilitate the entry of
means, it was possible to ind
profile within 3 months.

Preheaving by flooding ha

- The time necessary to ach
  of two to three months, ev
  into the soil. Careful plan:
  included at the start of the
- At the end of flooding the s
  immediate access to the si
  layer covered by a layer o
  frictional material must b
  stabilized soil can be consti
  has not become a popular

Figure 8.9 illustrates the cl
preheaving process.

In the virgin desiccated pr
Figure 8.9a). u_0 is in dynamic
surface. The corresponding po
(Figure 8.9b) would be A. At t
been established at the groun
Figure 8.9a). The surface has
B in Figure 8.9b describes the

_____
Permanent water table

Tension
(negative)

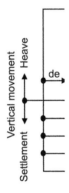

Vertical movement

Heave

de

Settlement

Figure 8.9 (a) Changes of pore pressu
            (b) Change of effective stre

When the soil surface is covere
of moisture which will gradua
with the permanent water tabl
in Figure 8.9a). The effective s
will have settled, bring the soil
loading imposed by the structu
stress further and cause further

Figure 8.10 shows the tim
de Wet, 1965) field experimen
period the surface of the test ar
Heave was observed to continue
menced. After 7.5 years the se

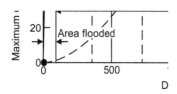

Figure 8.10 Heave dur

post-flooding settlement amou
post-flooding settlement will
deeper or the superimposed lo.
any deep-rooted vegetation w
lyptus or poplar trees or fast-gr
By re-desiccating the soil and c
have been disastrous for light
The broken curve superimpose
for a house constructed on the
seasonal accumulation of mois
the experimental time-heave cu
that can be brought about by i

The pre-heaving techniqu
shopping complex on alluvial a
South Africa (Blight *et al.*, 198
fill underlain by 1.5 m to 2 m
in turn, rests on 9 m to 10 m of
silt. This overlies less weathere
of 11 m.

The main structure was to
to stabilize the soil underlying
diameter and 6 m deep were dri
to main column grid-lines. Th
from collapsing when filled wi

Expansion of the soil wa
depth extensometers. The exte

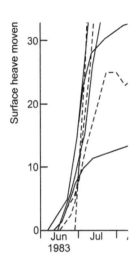

Figure 8.11 Time-surface heave relat
complex in Vereeniging, S

various depths in the profile to b
2½ months. Figure 8.11 shows
Figure 8.12 shows profiles of he
flooding was terminated.

The maximum recorded su
11 000 m² area of the site. Th
similar features to those of the t
the surface pegs were destroyed
were destroyed shortly thereafte
a steady settlement as the exces
to follow the continuing course

Figure 8.12 shows the varia
someters. These curves show tha
the profile where desiccation sho
ing has been carried out on the st
have been experienced and that

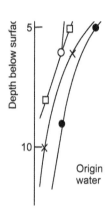

Figure 8.12 Heave-depth relationshi
South Africa.

Blight (1984) described a
tension piles. The piles were al
in a profile of residual siltston
water table before flooding wɛ

The object of the experirr
piles as the profile heaved. Flo
installed on a 2.63 m grid and
were 75 mm in diameter and
hose pipe over its full depth. F
extensometer.

The pile group was first
the effects of this initial wettin
50 days.

Figure 8.13 shows a reco
and 14 m below surface. The c
watering system. There is a ver
and the heave curves. Unfortu
is not known with any precisi
the volume of water injected a
Approximate calculations sho
lost from the test area by later

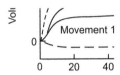

Figure 8.13 Relationships between t
experimental pile site.

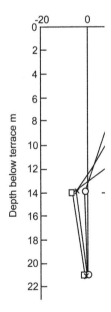

Figure 8.14 Heave-depth r

Figure 8.14 shows relation:
the start of flooding. The data a
and 8.12.

In this case, therefore, the te
of accelerating the development

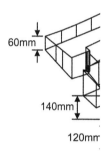

60mm

140mm

120mm

Figure 8.15 Effect of surface irr

Much of the heave at a pa
the soil profile, where desicca
that surface flooding or irriga
sufficient time is allowed for r
probably be at least 2 to 3 mc
gation as a remedial measure
storey load-bearing brick apar
soil profile that consists of 1 m
of desiccated residual shale. N
The apartment block consisted
being linked by walkways on c
surfaced with asphalt concrete
occurrence of heave.

Cracking of the buildings
was delayed because of this. W
and cracked that it was feared
complex, and level measuremer

Figure 8.16 Effect of surfac

and tilted outwards, moisture h
courtyard.

It was decided to attempt tc
spray irrigation. This was und
much reduced. Figure 8.15 shov
the complex, while Figure 8.16 :
after irrigation. However, it wil
tial movements were maintaine
Figure 8.15 and 90 mm betweer
tain the site in a heaved condit
open-jointed concrete blocks b
rainfall to penetrate into the soi
surface.

Ten years later the remedia
had been a slow settlement of tl

The case histories have sho·
be successfully carried out. Pre
careful monitoring of the effect:
the time taken for the water t
scheduling can overcome this sr

8.5 HEAVE ANALYSIS FC
EXPANSIVE CLAY A1

The observed time-heave record
has been shown in Figure 6.43
record covers the first seven yea
of the house appears to have er
movements and movements due
have occurred. The movement c
shown in Figure 8.17b. The hea
in what follows.

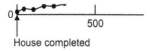

0

500

House completed

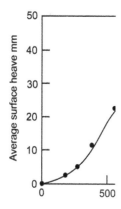

Average surface heave mm

50

40

30

20

10

0

0

500

Figure 8.17 Heave vs time curves fo
wide road adjacent to t

8.5.1 Similarities betwe

Settlement of a saturated clay
pressures that are out of stati
tlement proceeds by the expul
pressure beneath the settling s
table. The relationship betwee
saturated soil can be written a

$$\frac{\Delta h}{h} = C\,\Delta(\sigma_v - u_w)$$

where $\Delta h/h$ is the settlement s
        C is the compressibility
        σ_v is the vertical total :
        loading,
        u_w is the pore water pr

... soil, c_v, on the drainage con...
location of free surfaces and per

In order to perform a settle
settlement curve, it is therefore

- by means of laboratory tes
  ibility and coefficient of con
- to estimate the profile of e:
  and
- to assess the drainage condi

In a partly saturated soil p
suctions are in dynamic equilibri
prevailing on the surface. In th
moisture deficiency in the profi
equilibrium with the water tabl

When the surface of the soil
moisture deficiency is gradually
structure approaches static equi
under the structure, the soil sv
between the suction profiles for ¢
the excess negative pore pressur

The relationship between h
saturated soil can be written:

$$\frac{\Delta h}{h} = C\{\Delta\sigma_v - \Delta[\chi u_w]\}$$

In this equation:

$\Delta h/h$ is the heave strain, an

χ is the Bishop effective str
on the pore water pressure u_w.

Once again, σ_v will remain virt
can be rewritten as:

$$\frac{\Delta h}{h} = -C\Delta[\chi u_w]$$

depend on the dimensions of t
swelling process.

The heave and settlement
differences arise from differenc
soils.

8.5.2 The profile of exc

In the case being considered, ir
sive clay and also to estimate
block samples were taken from
house and an adjacent asphalt
diameter were subsequently tri
pressure in a triaxial cell, and
axis translation technique. The

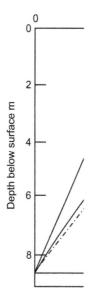

Figure 8.18 Measured, calcı

In the absence of better da
commencement of heave lay sor
The assumed suction profile is s

The excess suction for heav
suction profile and the line of h

8.5.3 Measurement of th
for diffusional flow

A clear distinction must be ma
tion. In the former process the s
large voids in the soil and flow
soil. In rain water penetration, a
through fissures in the soil. Pen
ably more rapid rate than diffus
to be due to diffusional flow. I
negligible time.

The coefficient of swell for
measured on samples from dep
over-burden pressure in a triaxi
A controlled pore water pressur
decrement of suction and the ra
of movement of a mercury threa
intake volume versus time curve
value of the coefficient of swell
the observed intake-time curve t
theory at a fractional intake of

Figure 8.19 shows the mea
various suctions.

8.5.4 Drainage conditio

The most difficult problem is to
for the heave movement to take
drains, there are three possible s

- upward diffusional flow of
of 9 m,

0
0   0.5

Corrected

Figure 8.19 Relationships between s

- penetration of rain-water (
  along the perimeter of th
  the soil beneath the struc
  minimum wetting path ler
- lateral penetration of rain
  soil, followed by downwaı
  path length of 9 m.)

It appears likely that the h
ture from all three of these so
the effects of one or two of the
   The relative importance (
comparing the observed beha
assumption of different draina
   Figure 8.20 shows a smooı
and external pegs in the experiı
shows the variation of the per(

$$T = \frac{c_s t}{D^2}$$

where c_s is the coefficient of s
   t is the time after comp
   D is the length of the w

c_s has been taken as the mea:
average suction in the soil pro:
   For a constant isotropic c_z
and hence lateral diffusional fl
either form of vertical flow (D

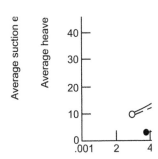

Figure 8.20 Comparison between o
            predicted dimensionless ‹

For the 6.1 m wide road,
diffusion would also be the mos

Upward diffusional flow froι

The curve, if it is assumed that
the water table only, is shown
time factor corresponding to 5(
observed time factor. Hence upˑ
factor in the heave process for t

Vertical rainfall penetration

An examination of the movem
house showed that seasonal va
It is therefore probable that raiɪ
Let it be assumed that the
the suction in the profile arouɪ
and that thereafter this conditɪ
calculated on this assumption ɪ
curve, if heave by combined hoɪ
is assumed. There is little differ
observed and calculated time fa

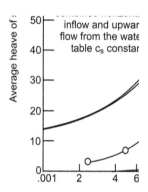

Figure 8.21 Comparison between calculated dimensionles

If the measured values of c
heave fits the observed facts su

Lateral rainfall penetratior

This mechanism of moisture ac
of the experimental house, wh
pervious surface stratum. In th
is more likely (but with a muc
    It can thus be concluded t
suggested, that described by t
important, although the influ
resultant time-heave relation.
    It must be stressed that th
consideration. With different
depths a different conclusion r

8.5.5 The relationship l

It was seen earlier that no direc
partly saturated clay and corre
these two properties was giver

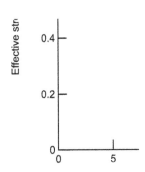

Figure 8.22 Relation between the efl
saturated expansive clay.

The relationship between tl
swelling of the clay is shown in F
decreases, i.e., as the heave pro
when the average equalization o
less than 50 per cent of the heav

It can be shown from equat
total surface heave which has ta

$$U_h = \frac{(\chi p'')_0 - \chi p''}{(\chi p'')_0 - (\chi p'')_{100}}$$

where $(\chi p'')_0$ is the average val
$(\chi p'')_{100}$ is the average v.

Equation 8.6 shows that th
less than that for suction equall
The steps in the calculation

- The observed suction equal
  calculate the curve of pore
  shapes of the experimental e
  it was sufficiently accurate,
- The resulting curve is correc

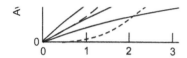

Figure 8.23 Comparison between c
house.

The predicted heave curv
values for the whole depth of
assumptions with regard to d
accuracy of the prediction, th
using average soil properties fo
the recalculation are shown in l
curve for the 3.7 m depth poin

8.5.6 Accuracy of the ti

The predicted time-heave curv
for the house and the road. I
observed movements and late
fairly close.
The initial lead of the pre
establishment of the assumed
latter stages of the heave may
extraneous sources such as a
allowed around the experimen
The accuracy of the time-l
shows the error in the time pre
average properties for the wh
for both the house and the roa
dimension but have similar wa
Using a two-layer calculati
materially and the accuracy of
different from that for the con

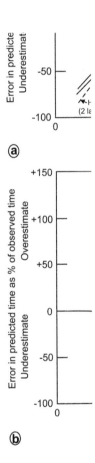

Figure 8.24 (a) Error diagrams for pr(
Error diagrams for predi(
clay (after Skempton & Bj

To put the error lines o
Figure 8.24b which shows sim
yses of three structures founde(
clays (Skempton & Bjerrum, 1⁹
are of the same order as those iḻ

concrete piles. The piles were
less weathered dimensionally :
weathered siltstone. All struct
between the soil and the under
would not be problematic, as t
Lightly loaded piles would ha
the uplift forces imposed on th
apply to piles that ultimately
ground for long enough to be
were applied.

There were three basic pro
Collins (1953) had propo
expansive clays. This method l
out ever having been checked
Donaldson (1967) had carried
mented pile of 230 mm diame
form of Collins' expression foi

$$P = \pi D \int_0^L (c' + K\sigma'_v \tan \varphi'$$

In which P is the tension in th
D is the pile shaft di.
L is the length of
anchorage,
c' and φ' have their t

However there was uncert
very large piles that would be
diameter):

- There was uncertainty as t
- What value of K should b

Donaldson's study had n
tensions predicted from measu

were of 1050 mm diameter and
ure 8.25. The plugs were instal
soaked by filling the hole with v
to drain away until there was
Figure 8.25). This ensured that
was close to the value of the to
zero. The plugs were then pulle
the plug were fastened.

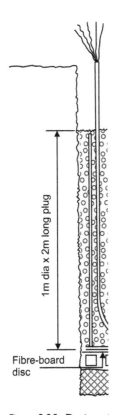

Fibre-board
disc

Figure 8.25 Design of

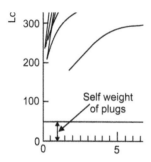

Figure 8.26 Load-displacen

Figure 8.26 shows the re
between the two plugs centre
other after soaking. Also note
full resistance.

Figure 8.27 summarizes
strength tests on the siltstone,
With the exception of one tes
with the lower limit to the lat
expectations, in the light of p
the test had been performed ai
oratory shear tests correspond
the need to know values for ε
In fact, the indication was thai

Design of piles

The piles could be designed u
changing effective stress and tl
laboratory shear tests had sho·
with one exception, the field c
strengths derived from the plu
had a considerable effect on th

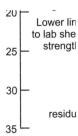

20
Lower lin
to lab she
strengt
25
30
residu
35

Figure 8.27 Summary of laborat

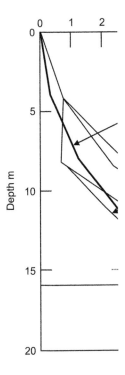

0   1   2
0

5

Depth m

10

15

....

20

Figure 8.28 Pile design curves based
progression of heave with

Figure 8.

8.6.2 Field test on an i

The one question remaining w
the distribution of tension in th
forces could correctly be pre
the question, it was decided t
be subjected to accelerated sv
on a 2.625×2.625 m grid, as
1050 mm in diameter and 33 m
instrumented piles were select
(2), and an interior pile (3) in ;
the reader is referred to Blight (
and vibrating wire – were usec
The soil surrounding the te
and the strain at various depth:
up water and swelled. (See Fig
Figure 8.30 shows the var
(see position in Figure 8.29).
the assumption that the concre
was carrying the entire load. S
not sufficient to crack the con
either end of the pile.
The design tension curve ·
Collins' equation (Equation 8.
sonable agreement with the m
underestimation at depths less

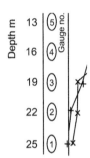

Figure 8.30 Development of tension i
depth-tension relationshi

Figure 8.31 shows the vari;
The diagram shows how uplift
counteracted by downward anc
diagram also provides a check
area under the uplift shear curv
vertical equilibrium is to be pres
downward) equaled just more tl
1500 kN. The difference betwee
because of their position in the
the measured shaft extension to
differed by 15%, with the strain
figure. The strength versus dep
has been superimposed on the u
the shear stress-depth curves cal
shaft shear stresses were undere

8.6.3 Effect of loading a

As many of the piles would be
was decided to study the effect c
to two of the instrumented pile;
compressive load was applied b
jacking off the anchors. As had

2

Depth m

Figure 8.31 Shear stresses develop
strength vs depth relatic

on the pile shafts and was com
shafts below this depth were u

8.6.4 Conclusions

The programme of field testin;

(i) The Collins expression fc
piles used at Lethabo.
(ii) Appropriate design para

In the past 30 years, the p
Problems were experienced wl
the supporting piles were open
of the duct and the soil had ca
a thunderstorm before the sid

Section 1.9 briefly dealt with pr(
illustrated by Plate C18, of a ca
side of the outlet conduit. This
are probably more likely to occ(
fill against soil to concrete inter
a role in the mechanism of failu
arching can by itself result in h

8.7.1 Gennaiyama and C

This case history (Ngambi *et al*
measurements made on the top
through earthfill dams, the Gen

Both dams are relatively l
dam has a box-shaped condu
tlement cross-arms were inst;
$k = 10^{-7}$ cm/s $= 0.3$ m/year) con
and volcanic ash (25% by vol(
weight of 18 kN/m^3. The Goi d;
mounted as shown in Figure 8.3
of three equal parts by volume (
ered mudstone and crushed unv
of 16.7 kN/m^3. The permeabilit

Figure 8.34 shows the settl
measured by means of the cros
of the conduit) the reduced sett
structure, even though it had b(
it in a trench 2 m below the lev(

Figure 8.35 shows records (
various pressure cells at the tw(
ments of the two dams were co
dam, the final overburden press(
pressures recorded on the horiz(
started, were 280 (E − 4), 290
nearly four times overburden, il

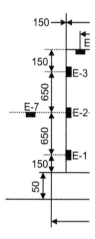

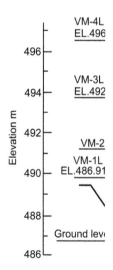

Figure 8.32 Details of Gennaiyama
conduit showing positi‹
settlement.

$\{$

$[$

Ground level

Figure 8.33 Cross-section of outl

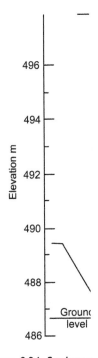

Ground
level

Figure 8.34 Settlemer

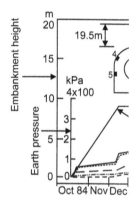

Figure 8.35 Pressures measured o
(b) Goi dam.

which behaved as if it had a
E − 1, E − 2 and E − 3, on the
and were very similar to the ve
in the backfill away from the s
was subject to a near–isotropic
was not measured.) Thus, a po
(30 to 40 kPa) would have red

$$\frac{\sigma_h}{\gamma_w h} < 1$$

a danger of hydraulic fracture
of Gennaiyama, both horizont
to the conduit and a distinct
failure by piping, must have ex
    For Goi (Figure 8.35b), the
ing for the crown of the cond

failure after hydraulic fracture. .
clay cut-off trench for the Lesa
concern that hydraulic fracturin

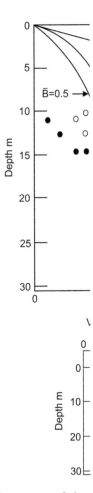

Figure 8.36 Comparison of obse

Vertical stre

Figure 8.37 Measured relationship t
trench at Lesapi dam.

section of the trench which w
of pressure cells and piezomet
along the length of the cut-off t
a compacted sandy clay residu

The three pressure cells w
and each pressure cell was acc
vertical pressure profiles were
vertical pressure ratio) of 2, w
vane tests in the adjacent com

Figure 8.37 is a plot of m
vertical stress recorded by the p
This shows that each incremen
increment of vertical stress an
As the vertical stresses, althou
much less than overburden st
horizontal stresses, as a result
stresses, it would not have bee
the core and hydraulic fractur

In the 40 years since th
attributable to excessive seepa

8.7.3 Concrete spillwa
concrete interfac

The Acton Valley dam is a rel
14 m. It is constructed of com

of water at the most.

Figure 8.38 shows a recons
AA shows the main flow path
formed simultaneously or seque

The first point of entry was u
crest slab about 12 m from the
of this slab and towards the tra
entry point. The second point o
the training wall. A whirlpool a
to the failure. (For clarity of t
Plate 8.1.)

The two flow channels app
shown in plan B-B in Figure 8.
behind and against the spillwa
adjacent to the training wall is
erosion cavity downstream of t
wall and the underside of the ir
daylight at the far end of the sp
a low head indicates that the w
loose uncompacted soil.

There are several likely pos
existed the failure.

Figure 8.38 shows that the
the dam [marked (CL) for clay
rests on the top of the filter drain
[marked (SM)]. The clay core v
flank it, because clayey soils ten
detail at the junction of the spill
this differential settlement, henc
crest slab as the clay core settle
flow from point of entry 1 in Fi

Apart from the visual evide
slab in 1988 and 1989 confirme
crest slab as a result of unaccon

The L drain shown on the dr
any small leak, such as the in
However, the drain did not fulfi

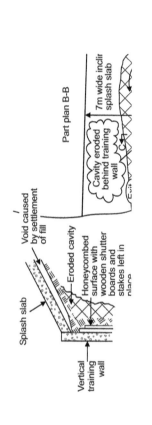

Splash slab

Void caused
by settlement
of fill

Eroded cavity

Honeycombed
surface with
wooden shutter
boards and
stakes left in
place

Vertical
training
wall

Part plan B-B

Cavity eroded
behind training
wall

7m wide inclir
splash slab

Exit

C

Plate 8.1 Entry ┌

initiated the leak allowed the fl
entry of water to the drain had
by grout during concreting the ┌

In the area behind the trainiⁱ
as the flow paths bypassed the dr
of the drain.

The height of the dam alon
by 4.6 m where the spillway cr┌
occurred between the spillway ┌
of the spillway where the failu
developed between the end of t
line with point of entry 1) by Se┌
hogged the crest of the spillwa⁻
spillway crest slab, and possible

In addition to the dragging ┌
lower ends of the training walls

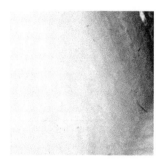

Plate 8.2 (View from right). Oı

the training walls. There is no
and the training wall to take ı
have formed under the splash ₁
illustrated in section C-C of Fi

After inspecting the dam ₁
differential settlement between
way, the first author telephonε
undiscovered, would probabl)
the spillway. On investigation,
resulted in a void under the spl;
occurred on the other side of t

Section CC depicts the dif
side of the splash slab and the ł
void that is thought to have oc
where the main piping failure

The erosion void behind t
was honeycombed in places aɪ
had been left in place behind tł
in view of this carelessness, ε
against the wall was not well
concrete surface. Hence the diٍ
splash slab was probably riddlɾ
compacted fill against the bacl

to recognize the importance of
which fill would be compacted
surfaces. Neither preparation fc
paction against concrete surfac
specification.

This case history had a sad e
to the dam and the crops lost be
on under a huge debt burden,
which his family had owned an

8.8 THE STABILITY OF S

8.8.1 General comment
of slopes in residu

A natural slope in residual soil
has, over millennia, removed w
and deposited it on the lower sl
stones has occurred, minor slur
Figure 8.39 shows measuremer
very ancient, 4200 million year-
hued talc content). The slope i
measurements show, the surfac
profile, with the upper slopes be
transported from above, gradua
rates approaching 50 T/ha/y at t
$(1\,T = 10\,kN)$. The lower part is
under the erosion rate curve (ir
tion rate curve, but erosion and
difference of 150 Tm^2/ha/y in 1
occurring from the slope abov
the accretion. Eventually, the up
cally different profile, and the e
established upper slope, will cor

The point is that natural s
disturbed by more than a small a

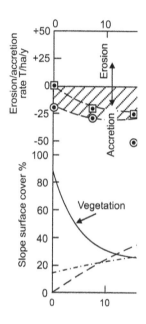

Figure 8.39 Erosion of a natural resi
of erosion/accretion, (c)

cover of editions 1 and 2 of this
cut-to-fill platform had been co
The disturbance was sufficient
a short wet period. After build
the house was completed, and
quite possible that the next ex

Plate C28, taken in 1963,
slope of residual mudstone (se
the slope which has a sea view
of cracked and repeatedly repa

... ...
weathered soil beneath. This m:
effect of the interparticle bonds
dilation of the soil, are gradua
Figure 7.11), the stability decre:
without flattening the whole slc

Another example of long-t(
a failure in a steep (1:1) railwa:
been made in the late 1920s an(
failures developed, one on each
heaved up the rail-bed, complet(
an attempt was made to effect a
with the even more disastrous (
section made after this disaster,
hastily constructed reinforced ":
was not the stabilizing buttress

8.8.2 The largest ever fa

In terms of volume of soil mobi
took place at Vajont in northe
had been constructed across a
Vaiont) river flowed. It was com
was 169 million m^3. In 1959 c(
slopes forming the sides of the
undertaken. The various analyst
as to the seriousness and extent
As a result, a system of monitc
and it was decided to carefull
measurements of surface moven
that the limiting amounts and r
be identified.

On 9 October, 1963, the sol
moving at a speed estimated at a
weathered rock of 275 million
140 m above the full supply level
reservoir, it displaced the water :

Figure 8.40 Vajont slide – section ac

over the crest of the dam. The c
five villages downstream, killii
reservoir, indicating the main s

The next biggest slope fai
that was triggered by the cutti
of the Panama canal (McCulle
through the highest point of t
25.9 m AMSL. Excavation of t
Canal Interoceanique, fronted
Engineers gave problems from
the canal is being widened. Th
shale, the Cucaracha shale, wh
debilitating cost. Plate 8.3, pho
shovel almost buried in slide d
the cut, the extent to which the
before the Canal company wer
excavation by the US before th

The lower part of Figure
cessively sloughed away. The
cut had flattened 33 years late
An American who worked on
following reminiscences (McC

"The whole top of the hi
composed of a clay that is utte
gets it on. He has to have a lit
kept the hill from sliding. "It v

Why? he was asked.

"The rainy season will sat
"Did it do so while you w
"Yes, we had a cut right a
it sloughed off, not only over
so expensive to move it that I

Plate 8.3 Steam shovel half-l

year or so afterwards the same
the present track is, there are tw

"... when I was there at C
four hundred to five hundred fe
out and the land had gone off a
not say anything, but I knew wh
down into the canal. Every rain
saturated and it slides right off

"It slides on the blue clay?"

"It slides on the blue [Cuca

8.8.3 Specific comment:
of instability

The effects of unusually hea

The effects of unusually wet w
been examined by many resea

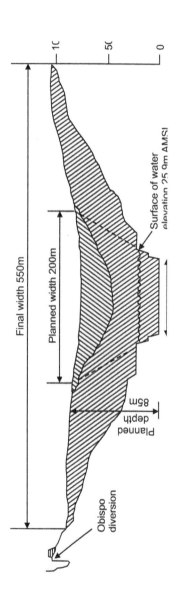

Obispo
diversion

Planned
depth
85m

Surface of water
elevation 25.9m AMSL

Planned width 200m

Final width 550m

10

50

0

geneous soil in the absence of c

permeability of the soil. The w
travels at an approximate veloc
initial degree of saturation and n
and rainfall intensity, Lumb sho
critical depth in a slope (such as
Open cracks and fissures in a so
wetting front.

One of the most extensive
which occurred in the Serra da
1969). During and following a
70 mm/hour, an area 24 km lc
landslides that killed an estimate

Van Schalkwyk & Thomas
KwaZulu Natal province of Sou
800 mm fell during a period of
to property and infrastructure c
this period, 211 slope failures c
In almost every case, the failure:
been subject to man's interferen
or railway alignments.

Effects of seismic events

Yamanouchi & Murata (1973)
residual volcanic soil) that occur
relatively rigid, brittle material,
cracks in the soil which caused s
appear to have been affected, cu

Human interference

There are many types of huma
slopes in residual soil. Of these

- Removal of toe support by
  introduction of a cut at the
  The slide that occurred at B

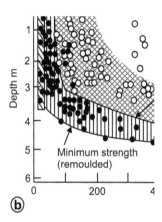

(b)

Figure 8.42 (a) Sectional block diagra
measurements in weath

example of this type of failt
road cut was made in a h
was found that the slide in
on its contact with an und
failure were exacerbated b
mudstone contact of a co
out of the sandstone layer.
    Material removed from
stable slope to become uns
converse effect caused by e
• Changes in the soil water
    If the soil water regime c
of vegetation or partial ir
(Richards, 1985). The mo
by raised water levels in a
valley described in section
• The effect of deforestatio
following appear to be the
rooted vegetation is remo

Many of the agencies that pro[...]
seismic activity, toe erosion, cha[...]
of cut slopes especially in the l[...]
to be dominated by the structur[...]

Residual soils that do not [...]
jointing, bedding or slickensidi[...]
slopes. For instance, Wesley (19[...]
a latosol clay. These slopes hav[...]
(see Figure 8.43). He also menti[...]
has been stable for over 40 yea[...]
stable vertical cuts of 5 to 6 m i[...]

Woo *et al.* (1982) list cut a[...]
west Taiwan. These gravels ar[...]

Figure 8.43 Cross secti[...]

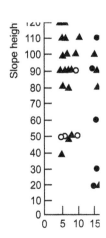

Figure 8.44 Slope height vs angle fc

in situ. Woo *et al.*'s observatio
ably have angles of shearing r(
lateritization, it is only the ca
that may be unexpectedly stee:

The stability of the slopes
by invoking the effects of por(
vertical slope with the water ta
if the capillary stresses within
This applies whether or not th
how many such slopes remain

Slides caused by the prese
difficult to predict or design a{
hole specimens and may be d
trenches. Even if potentially da
exploration, it is usually practi
tion in dip and other factors th
1985).

St. John *et al.* (1969), n
attributable to the presence of s

bounded by intersecting and detail

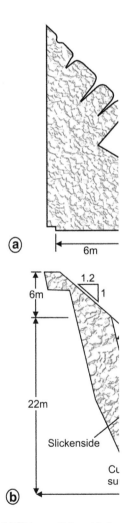

@a ⊢───────⊣ 6m

1.2
1

6m

22m

Slickenside

Cu
su

ⓑ ◄───────

Figure 8.45 (a) Slide on slickensided s
1969). (b) Slide on inters
tuff (after St. John *et al.*, I

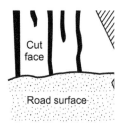

Figure 8.46 Plan showing three di
Maurenbrecher, 1974).

Another example has be
Figure 8.46) in which a wedge
diabase. The failure occurred
and the pre-existing sliding sur
of a hornblende-like secondar
this failure is clearly shown by
almost intact wedges of soil th
Figure 8.46.

8.8.4 Design of slopes ₂

Because of the unknown and p₍
discontinuities in a residual soi
basis. For this reason, many b
ported, the support system bei
(e.g., Flintoff & Cowland, 19₈
unsupported, but the economi₍
vastly more serious than the c₍
such as road or rail cuts, slop
ad hoc remedial measures are
As the surfaces exposed by a s
than the original cut face (e.g.
clearing away the fallen debris
In other cases drainage m
provision of toe support in th
Nunes (1969) describes the ₎

Reinforced
waler beam

A
b

Figure 8.47 (a) Slip in a cut in residua
anchors (after Wagener &

Gabion or reinforced earth wall
essary to tie back the slope usin
A section through this potential
in Figure 8.47.

If rational design and analys
a decision has to be made as to
value may be used in the design
in which it is assumed that eithe

(a) stability is controlled by t
     failure surface,
     or
(b) stability is controlled by
     tial failure surface. This
     soil behaviour should lie.
     strengths are established in

There is, however, consider
that the shear strength of a stiff f
of strengths measured in small-s
8.41b and 7.41b). This was als
back analysing failures of four
and in relation to series of slide

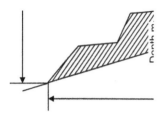

Figure 8.48 Detailed section throug

mudstone at Amsterdamhoek,
the slide and a typical vane stre
the shear strengths, back-calc
as a strength versus effective s
reversed shear box tests done
apply to three residual soils in I
Shelton (1962). The data are sl
will be noted that in each case t
(triaxial shear) corresponds to
The analyses assumed zero p
probably a small suction pres
has probably been under-estir
the laboratory had been used t
result would have been obtain

8.8.5 Types of failure o

Many attempts have been ma
Classifications depend on twc
sliding soil mass and the veloc

Geometry of slides

There are two main types of
in comparison with the lengtl

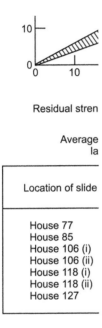

10

0

0   10

Residual stren

Average
la

Location of slide

House 77
House 85
House 106 (i)
House 106 (ii)
House 118 (i)
House 118 (ii)
House 127

Figure 8.49 Shear strengths back-calc

the soil mantle slides over the
mass is usually roughly constar
vex or slightly concave. The C
(see Figure 8.51) and the slid
cal planar slides. Other typical
slopes have been those at Bethl
Figure 8.47.

The second typical slide geo
a rotational or gouging slip or a
described by Morgenstern & de
while the Danube slide of a rive
block gliding failure (Figure 8.5

Figure 8.50 Comparison of lower lir
from slope failures in th

Slide velocity

The velocity of a land-slide ma
of several kilometres per hou:
afforested slopes in Oregon l
on the southern California co
100 mm/year, and one in the C
ities are by no means constant
by wet weather and retarded b

Figure 8.51 Section through a typical
Pichler, 1957).

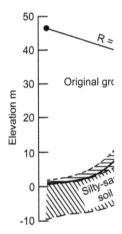

Figure 8.52 Section through a typical
Morgenstern and de Mat

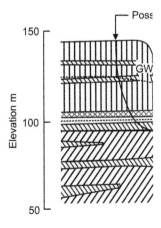

Figure 8.53 Section throug

Annandale, G.W. (1979) Settleme
 Soil. MSc(Eng) Dissertation, Ui

Barksdale, R.D., Bachus, R.C. &
 Eng. & Const. Tropical & Re
 USA. pp. 647–664.

Blight, G.E. (1969) Foundation f
 Div., ASCE, 95 (SM3), 743–76

Blight, G.E. (1973) Stresses in narı
 Large Dams, Madrid, Spain. pı

Blight, G.E. (1980) Partial saturat
 Taipei, Taiwan. Vol. 1, pp. 15–:

Blight, G.E. (1984) Uplift forces ı
 Soils, Adelaide, Australia. Vol.

Blight, G.E. (1987) Lowering of ı
 Soil Mech. & Found. Eng., Dul

Blight, G.E. & de Wet, J.A. (1965
 ria and Moisture Changes in Sı
 pp. 89–92.

Blight, G.E., Brackley, I.J. & vı
 Bethlehem – an examination of
 June, 129–140.

Blight, G.E., Legge, G.H.H. & An
 earth embankment. Civ. Eng. Sı

Blight, G.E., Schwartz, K., Webeı
 flooding – failures and successe

Brand, E.W. (1982) Analysis and
 Soils, ASCE, Geotech. Div. Spe

Brand, E.W. & Phillipson, H.B. ı
 Scorpion.

Burland, J.B. Butler, F.G.B. & Du
 bored piles in stiff clay. Symp. I

Collins, L.E. (1953) A preliminarı
 South African Inst. Civ. Engrs.,

Da Costa Nunes, A.J. (1969) Lanı
 7th Int. Conf. on Soil Mech. &

Donaldson, G.W. (1967) The me
 Soil Mech. & Found. Eng., Capı

Flintoff, W.T. & Cowland, J.W., ı
 & Construction in Tropical &
 USA. pp. 539–556.

Vol. 1, pp. 566 571.

Lumb, P. (1975) Slope failures in H

Malone, A.W. & Shelton, J.C. (19
Construction in Tropical & Res
USA. pp. 424–442.

McCullough, D. (1977) The Path I
1914. New York, Simon & Schu

Mitchell, J.K. (1976) Fundamental:

Morgenstern, N.R. & de Matos, M
Conf. Soil Mech. & Found. Eng.

Muller, L. (1964) The rockslide in
148–212.

Muller, L. (1987) The Vajont catast

Ng'ambi, S.C., Shimizu, H., Nish
ure mechanism of embankment
Blight, G.E. & Fourie, A.B. (eds
Balkema. pp. 647–652.

Pavlakis, M. (1983) Predictions of .
Tests. PhD Thesis, University of t

Pells, P.J.N. & Maurenbrecher, P.M
South Africa, 16 (5), 180–181.

Popescu, M. (1998) Engineering pe
lower course. UNSAT '98, Beijin

Richards, B.G. (1985) Geotechnica
Phillipson, H.B. (eds) Sampling
pp. 23–30.

Sandroni, S.S. (1985) Sampling a
Phillipson, H.B. (eds) Sampling
pp. 31–50.

Skempton, A.W. & Bjerrum, L. (19
on clay. Géotechnique, 7, 168–1'

St. John, B.J., Sowers, G.F. & We
engineering significance. 7th Int.
Vol. 2, pp. 126–130.

Ter-Stepanian, G. (1963) On the Lc
No. 52, Oslo, pp. 1–13.

van Schalkwyk, A. & Thomas, I
September 1987 and February 1
Wardle, J. & Fourie, A.B. (eds), (
pp. 57–64.

Spec. Conf. Honolulu, USA. pp

Yamanouchi, T. & Murata, H. (1
   Conf. Soil Mech. & Found. Eng

Yen, B.C. (1969) Stability of slope
   ASCE, 95 (SM4), 1075–1096.

activity
    biotic 157–159
    Skempton's 5
African erosion surface, Above Afr
    Post African 9
air
    displacement of water from soil l
      121–122
    permeability, measurement in lab
      130–133
    pore pressure, undrained compre
      111–115
    steady state flow, Fick's law 119
    unsteady flow of, through unsatu
      and dry soils 122–125
alkali, alkali earth 9
alucretes 14
aluminium sesquioxides 14
amygdales, vesicles Plate C2, 13
analysis of heave, desiccated expan
    301–312
andesite, composition, mineralogy
    13, 27–29
anisotropic stress and/or strength 2
apartment block, collapse settlemer
area ratio (C_a) 77
ash, volcanic 3
at rest in situ stress ratio (K_o) 253-
Atterberg limits
    duration and method of mixing
      53–55
    pre-test drying 52–53
augite 27

bacillus, thio-bacillus 4
bacteria, bacterial oxidation 3, 4
bentonite, montmorillonite, smectit
biotic activity 157–159

of 43
compressibility (C)
  collapse of compacted soils 10?
  measuring 156–178
  weathered andesite 29, 65, 149
consolidation
  coefficient of (c_v) 176
  time factor (T) 176
constant
  effective stress state 154–155
  or falling head permeability tes·
    127–145
core boulders, stones Plates C3, (
cretaceous era 9
crumb test, dispersive soils 25
crystallization and weathering or(
    minerals 48–49

dam, earth embankment, settleme
    289–293
Darcy's law, steady state flow 11!
definitions, residual soils 1–3
degree of saturation (S) 29
delta deposits 1
density, dry, maximum 81
deposition of salts 1, 2
depression of water table, evapot·
    192–194
depth of
  weathering 17–18
  seasonal movement 192–194
description, profile 65
desiccated clay, heave analysis 30
diffusion coefficient (D_c) 119
discontinuities, effect on strength
displacement of water from soil b
    121–122
dispersive soils, crumb and pinho·
    25–26, 47

flow, steady state, water, air, Darcy
    Fick's law 119–121
fluvial deposits 1
Fredlund & Morgenstern's, ϕ^b 218

general characteristics, residual soil
gneiss 15, 16
Gondwana, Gondwanaland 9
gradient, pressure (i) 119

halloysite 9
heave of desiccated expansive clay
    analysis 301–312
    distribution with depth, 192, 19
      299
    prediction of 194–201
    uplift on piles 193–194, 298–29
      314–318
    variability 300–301
heave predictions compared with s
    predictions 311
Henry's coefficient of solubility of ε
    water 113–114
Hilf's equation 113
holes, pits, trenches for soil explora
    63–64
hydrolysis 3
hysteresis 35–37

igneous rocks 3
illite 3
index parameters, difficulties in
    measuring 50
infiltration (ponding) tests 133-13
infiltrometer, single & double ring
influence diagram, strain 179–183
in situ tests
    shape factor (F) 137–144
    soil sampling 73
    water permeability 142-145

soils 166-201

$m = \sqrt{(k_h/k_v)}$ 139, 141

N, standard penetration test num
    175–177
N_c, bearing capacity factor 270, :

oedometer test, double 190–192,
onion skin weathering Plate C1,
optical microscopy (OM) 43
optimum water content, maximur
    density, saturated water perm
    93–94
origin and formation, residual soi
oven drying, effects of 25, 50–52
overconsolidated characteristic, y
    149–153
oxidation, bacteriological 3

pan evaporation, A-pan 64
partial saturation, effect on streng
    218–222
particle size distribution (psd) 55
particle specific gravity, relative d
    G_s) 2, 35, 55
ped, intra-ped 41
pedocretes
    deposition, forms, rates of forn
        18–21
    foundations, road construction
        blocks, embankments 20–21
penetration test, standard (SPT,N
permeability
    air flow, measuring 130–133
    borehole inflow or outflow 137
    in situ testing 101, 132–145
    profile description 64–73
    saturated, water flow 64–65, 9
    small scale and large scale meas
        differences 125–127

permeability characteristics 145–
relict structures
   stability of slopes 329–346
ring shear test 232–235
rock, definition 1
   pinnacles plates C15, C16, 21–2
   weathering 3–4
roller compaction, field 90–93

salt deposition 18
   roads Plate C23, 35
sampling, soil – firm, soft and stiff
   silts 73–79
saprolites, saprolitic soils Plate C6
saprolitic structure, effect on streng
   222–225
saturated, permeability, water flow
saturation, degree of (S) 29
scanning electron microscopy (SEM
sedimentary rocks, weathering Plat
   C12, 17
sedimentary, sediment 3
seepage, compacted clay, field com
   94–96
seismicity 64
sequences, weathering 11
sesquioxides, aluminium, iron 1, 2
settlement (ρ)
   apartment block collapse, loess
   compacted earth embankment 2
   tower blocks, residual andesite 2
settlement of foundations, Menard
   185–186, 286–288
settlement predictions
   collapse, residual soils 201–210
   deep foundations 186–188
   raft and spread foundations 178
shallow foundations, movement, re
   soils 188–201

test holes, pits, trenches for soil c
    63–64
testing, laboratory 74
test pad, large-scale water permea
    144–145
thermo-gravimetry (TG) 43
thio-bacillus 4
time-factor (T), consolidation 17
topographic relief, landform 8–1
total dissolved solids (TDS), dispe
    soils 25
tower blocks, settlement, residual
    286–289
transmission electron microscopy
transported soils – weathering in
    sands, volcanic ash, Plate C1
triaxial test
   apparatus, test samples and var
    235–244
   consolidation and drainage stat
   loading (deviator) stresses 241
   measuring K_o 253–257
   multi-stage tests 244–246
   pore pressure equalization, rate
    247–250
   pore pressure measurement 24
   saturation of specimen, back pr
    242–244
   testing cell 237–238
   testing stiff, fissured soils 250–

undisturbed soil sampling 74–79
undrained compression, pore air
    111–115
united soil classification system (l
    56–58
unsatisfactory, compaction, conse
    84–85

C.1 (View from right)

C.2 (View from right) Dε

C.3

C.4 Residual weathe

C.5

C.6 Profile of res

C.7 Lateritic grave

C.8 Layer of calcr

C.9 Profile of dole

C.10 Profile of

C.11 Relative

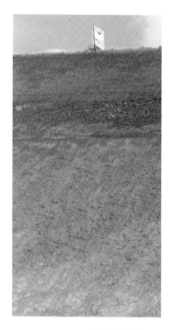

C.12 Shale pro

C.13 Colour

C.14 River course that has cut throu
ash by a recent eruption.

C.15 Near-surface rock p

C.16 Damaged buildin

C.17 "Bad-l

C.18 Piping fa

C.19 Extensive slickenside

C.20 Slickenside

C.21 Rapid v

C.22 Roman aqueduct in Toledo, Spa

C.23 "Salt road

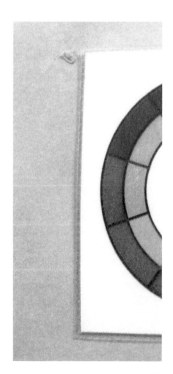

C.2

C.25 Example of variability

C.26 Post-construction underminin
weathered mudstone.

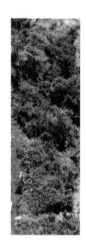

C.27 Shallow slid

C.28 One of a serie

C.29 Failure c

C.30 Slide in

C.31 Abortive atten

C.32 Three-d

Printed and bound by CPI Group (UK) Ltd, Croydon, CR0 4YY

18/10/2024

01776249-0004